Tornadoes:

Be Informed – Be Safe

Joe R. Eagleman, PhD

Tornado Photographs on Front and Back Covers NOAA

ISBN: 9798218708344

CONTENTS

PREFACE

An understanding of tornadoes and severe weather should be a fundamental part of everyone's storehouse of knowledge. We live in a world that is at least occasionally dominated by severe weather. Many types of severe weather are sufficiently rare that a common defense mechanism of some people is to assume that they will never be directly affected. However, severe thunderstorms are so common with lightning and the possibility of tornadoes that they require some degree of understanding and preparedness in order to avoid loss of property and, perhaps, even life itself. Tornadoes have occurred in every state in the United States with about 1200 forming annually.

The reader is provided, in this book, with the basic information on the formation of severe thunderstorms and how tornadoes are generated along with information on dealing with them through building design as well as many other more personal preparedness options.

The aim of this book is to provide everyone with the basic essential information on storm development, tornado formation and safety.

1. PROPER RESPECT FOR THE TORNADO

It is no wonder that so many people have an unhealthy fear of tornadoes for they have been known to do many strange things. Early on the morning of March 3, 2020 an EF4 tornado hit the Cookeville, Baxter, Tennessee area where Jennifer Henson-Collins lived with her family. "I heard the siren go off and grabbed my kids and we took shelter in the laundry hallway. The tornado was in our back yard! The tornado picked us up with the hardwood floor and landed us 250 feet across the street! We lost everything! I could feel my guardian Angel right behind us hovering over us to protect us. We were hollering out to God for help and he helped us! We had minimum injuries. The scariest night of my life. We're still trying to figure things out. Our lives have been forever changed since that early morning tornado."

In Xenia, Ohio a man heard a knock at the door and got up to answer it. When he opened the door it was a tornado. He and his wife had time to seek shelter although others are not always so fortunate. The loud noise accompanying a tornado may be the first warning that a tornado is near.

In another tornado a man heard a loud noise and went to the door to see what it was. When he opened the door the door flew out of his hand and he followed along with it. His friend went to the door to see where the other man had gone and saw him flying over the treetops. The second man was also pulled out the door and picked up by the tornado. Both of them were carried about 200 feet. One unwrapped the other out of a bunch of wire and they crawled back to the house. A tornado has sufficient strength to lift very large objects and, therefore, it is not unusual for them to pick up people, animals or other things in their path.

In Mountain View, Missouri a tornado picked up a man and carried him up into the storm. While he was flying about within the tornado a car came floating by. He tried to grab the door to get inside but missed and was not able to catch it. The tornado set him down in a nearby field, anyway, without doing him any great harm.

In Kansas a man was picked up by a tornado and carried along for a time. He remembered that as he flew about, he grabbed at anything within his reach. After the tornado had set him down, he found that he was clutching a handful of coarse hair that could only have belonged to the tail of a horse that must of been up there with him.

Another tornado picked up a house from its foundation and blew a cow into the basement. It then set the house back down on the foundation. This left a major problem for the owners who had to get the live cow out of the basement after the tornado had passed. Cows are not usually accustomed to climbing stairs.

In Oklahoma a tornado sucked a man, who did not have his seat belt fastened, right out through the top of his convertible and set him down in a field some distance away. On another occasion, in Oklahoma, it picked up the whole car with the man still inside and set it down on the flat roof of a building. This building happened to have stairs leading down through the kitchen of a restaurant. The man was able to walk down the stairs through the restaurant saying to the employees, "I'll be back tomorrow to get my car." The people not knowing that there had been a tornado and knowing that he could only have come from the roof, thought the man was crazy.

However, as is shown in Figure 1-1 tornadoes do not always handle people or objects with tender loving care and definitely should be avoided. With wind speeds as high as 300 miles an hour even very small objects such as straws, gravel, or even small pieces of wood may have penetrating power much like a bullet.

Figure 1-1. This photograph shows the mystical as well as destructive forces of tornadic winds in a way not obvious at first glance at the scene. The splinters are the remains of a large, strong utility pole and the cloth appearing material wrapped around the pole is actually strips of metal. Two books, one without missing pages are left by the tornado in front of the pole. (Photograph by the author after an EF5 tornado)

Tornadoes have been known to lift a 120 ton steel bridge from its abutments and drop it into the river, to lift five Pullman coaches from railroad tracks, each weighing about 70 tons, and deposit them as far as 80 feet away. Automobiles are frequently picked up by tornadoes and may be banged against stationery objects at a very high velocity so that they can be completely crushed by a tornado.

There is no doubt that people living in cities that have been struck by tornadoes, react differently to threatening weather than those living in cities that have not had recent tornadoes. This may be one reason why there are more deaths in the Southern United States from tornadoes than anywhere else, even though there are more tornadoes in the central United States.

William Powell in, *An Introduction to Natural History of Disasters*, has identified the different stages that seem to correspond to individuals stress reactions. They are warning, threat, impact, inventory, rescue, remedy and recovery.

Warnings may be broadcast for an approaching tornado by radio or television stations. One of the first was in 1955 on May 25 when the Weather Bureau issued tornado warnings several hours prior to the destruction of the entire town of Udall, Kansas. Even though local stations broadcast these warnings, 75% of the victims denied getting any notice. Possibly the weather reports over the radio were too technical, but more likely the people were unaccustomed to receiving and responding to severe weather warnings since the first tornado forecasts were made in 1948. After the Udall tornado in 1955 warning procedures were clarified and improved considerably.

The next stage of **threat** begins with the perception of the local change indicating a danger and may continue into the impact stage. At this stage persons become aware of danger and react individually. Some persons may take cover, some may panic and others may go outside to observe.

With **impact** disaster strikes. Activities are sublimated to self-protection. A community has been invaded and persons in groups become isolated for a while. Regular community service is disrupted and controls disappear.

Next comes **inventory**. The first reaction is bewilderment. Many TV news accounts show bewildered individuals after tornadoes sifting through the remains of their home and property. A period of reorientation is needed for a return to normal behavior. This explains why things happen as in a Waco, Texas, tornado when a girl climbed through a broken display window, calmly laid the money down for a record and walked out again, even though the plate glass front of the building had blown out and articles were flying through the air outside the building. In Udall, Kansas, a boy who had been through four tornadoes, after attending his mother and other children, hitchhiked to a nearby town to inform police and get help. Many adults were completely unable to function.

The next stage, **rescue**, brings help from others who have been less affected by the disaster. Help comes first from nearby and then farther away. Rumors may spread that can hamper an effective rescue. If authorities of the community do not take over and make a real assessment of the damage, rescue attempts may be put in a state of confusion and areas of severe destruction may go unassisted.

The **remedy** stage brings real organization. In the Udall tornado it brought a release into almost cheerful behavior which has been called "fiesta at the disaster." Specialists take over such as the National Guard, FEMA and the Red Cross.

The final stage is **recovery** but a community struck by a severe tornado cannot return to its original state. People and homes are gone, scars left. In Udall, Kansas, all of the houses were destroyed. These were rebuilt as modern ranch style houses with the anticipation of becoming a commuter suburb for Wichita workers instead of the sleepy village that had been there before.

It is important how we react to tornadoes. Proper reactions are not fear and panic. Although we cannot currently control tornadoes, we can decrease their impact through proper understanding that can replace much of the mystery and extreme fear of tornadoes. A proper understanding of tornadoes and the way that they damage houses may also save your life by allowing you to seek the best available shelter or even to design or construct a house in such a way that it will be least likely to be severely damaged by a tornado.

People react differently when experiencing a tornado. Topeka, Kansas was supposed to be a tornado proof city. There was an Indian legend that said that Burnett's Mound on the Southwestern edge of the city would deflect any tornado funnels away from the city. But on June 8, 1966, a storm was reported to be approaching the city from the Southwest. Five squad cars from the police force were dispatched and ordered to proceed to vantage points. One officer proceeded to the top of Burnett's Mound. When he got to the crest he saw only a large flat cloud spreading out on the rolling farm land. He wondered if any tornado could be that big. When he radioed his last report at 7:02 p.m. he headed down the hill to Interstate 70. Here he began to halt traffic and instruct motorists to take cover under the underpass. This proved to be an excellent shelter. The people who stayed there were battered a bit by the tornado that struck at 7:14 p.m. but all escaped serious injury.

As reported by the Topeka Capitol-Journal at 7:00 p.m., a nineteen year old newscaster for WREN radio station, Rick Douglass, left and headed for Burnett's Mound. He did not know as he drove up the hill that a tornado was coming up the Southwest side. At 7:10 he radioed the station that he was beginning his climb to the summit of the hill. When he reached the top he realized he was meeting a tornado head-on. As he turned his car around to drive back to town he radioed to the station saying that he was witnessing a monster storm with a super tornado (Figure 1-2).

As he got down to where the officer was directing people to get under the underpass, he stopped the cruiser near the I-470 overpass, several blocks south of S.W. 29th and Gage, to call in a report to the radio station. He saw Topeka police officer Dave Hathaway flagging down cars and telling people to climb the incline of the overpass and huddle beneath the bridge with their backs against the concrete. Just then,

Figure 1-2. Topeka tornado, June 8, 1966. (Topeka Capitol-Journal)

Douglass and Hathaway both looked over at Burnett's Mound. A long, black column several blocks wide descended from a gigantic cloud. It was heading toward them. Debris was churning in the air. "It had picked up the ranch house I'd just passed and blew it up like confetti," he said. Douglass grasped the mic of the two-way radio. "I can make a report on a house being blown away," he relayed. "Stand by," the radio station replied. Hathaway yelled at Douglass to get out of the car and seek shelter under the bridge. Douglass, who weighed 270 pounds, dropped the mic and followed the police officer, bracing himself against the wind as he tried to run up the incline.

Hathaway made it to the overpass and sat beneath the concrete structure with his hands clasped over the back of his head. Douglass stepped in mud and slid downward. He tried to hang on to a pillar of the overpass, but the wind knocked him to the ground and sent him rolling down the incline.

Suddenly, Douglass realized he was floating in the air. His legs and feet dangled about a foot off the ground. His arms were stretched in front of him. The sleeves of his jacket were inflated, and he could feel the jacket splitting up his back and the sleeves peeling away. Sticks, mud, rocks and dirt pelted against his skin and clothing.

As quickly as he was suspended in midair, Douglass went limp and tumbled on the ground. He crawled to a pillar of the overpass, but before he could wrap his arms around it, the wind began pushing him along the ground. He tried to grasp for grass or mud or anything to anchor him.

Douglass saw Hathaway spinning down the incline on the seat of his pants. Over Douglass' shoulder, he witnessed a house slide off its foundation and quickly disappear into the sky. A four-door Pontiac, with its doors and trunk open, was flipping in the air above him. After the tornado passed, Douglass was lying about a block north of I-470. The high winds caused mud and dirt to penetrate beneath the surface of his skin, but he survived the tornado.

Minutes later residents of the Huntington Apartments hid in bathtubs, closets and under tables. There was only one reported death from this part of the city. As the tornado crossed the city of Topeka it struck the campus of Washburn University (Figure 1-3). Fifty people attending a recital in a stone building (McVicar) heard the sirens along with the roar of the funnel and ran downstairs for cover. At this time the Weather Service advice was to seek shelter in the southwest corner. In the basement someone yelled, "Get to the southwest corner." They became confused and ended up in the southeast corner of the building. When the storm hit, the southwest part of the building was filled with tons of stones and debris so that they could not possibly have survived there.

Another family nearby laid down on the floor of their station wagon in their garage basement and watched as their house floated away. Most people in Topeka reacted as they

Figure 1-3. McVicar chapel on the Washburn University campus where 50 people sought shelter from the tornado.

should have during the tornado since only seventeen people were killed instead of a much larger number with over one thousand houses damaged or destroyed by this tornado.

A Topeka resident who experienced this tornado was Don Christie who tells us of his own personal experience and observations." At approximately 5:00 p.m. the local TV station issued a severe weather statement something like, possibility of tornadoes, etc. But Topekans were not accustomed to paying attention to it. Before this tornado whenever a warning was given for tornadoes people just went about their business. There was no difference as to the activities that went on. It was just as if the warnings had never been sounded. Perhaps the people were too accustomed to hearing the warnings to be concerned. Quite often they sounded and nothing happened.

Believe me, after this tornado, Topekans made a complete change. On clear days when the Civil Defense tested the warning system people were heading for cover. At about 7:00 on the day of the tornado the sky was very imminent that we were in for a storm. As the sirens sounded and the TV station stated that a tornado was aground on Burnett's Mound and traveling in a northeasterly direction we realized that our house was right in its path. My family headed for the southwest corner of our house for we have no basement. Now tuned in to the radio we waited for further reports. It was perhaps five minutes before we heard any trace of the funnel. It started as a faint noise but grew and grew until it was almost deafening.

I can now swear to the fact that they do sound like freight trains coming at you. This one sounded as though twenty or thirty were heading our way. Hearing the thundering tornado come closer gave me a desolate feeling knowing it was coming, but yet there was nothing you could do, just wait. I recall listening to it grow louder and louder. It reached a peak in volume and then started to subside. At that point I ran to the window to see what was happening. It had missed our house by a hundred yards and was receding toward the northeast. I ran outside and watched it go across town. It was truly an exceptional experience."

This first person account provides insight into some of the general characteristics of tornadoes as well as reactions to a tornado before and after a city suffers major destruction from a tornado.

You can watch storm chasers video of the largest tornado ever recorded (2.6 mi. diameter) from birth to death at: https://www.youtube.com/watch?v=Q7X3fyId2U0

2. UNDERSTANDING THUNDERSTORMS

Bill Farmer gives this account of severe thunderstorm development near the town of Pratt, Kansas. "This particular day was no different than any of the other warm spring days as most of those days involved a tornado watch in some part of the state. It so happened that on this day we were in the middle of a very large tornado watch. I was staying with a friend and we both went outside to look at the dark approaching storm from the southwest. It was about 4:30 in the afternoon and one of the things that struck me as weird was that there were no small cumulus clouds. Just one immense cloud with several others off it in the distance. I remember that a cold wind started blowing from the southwest. It never did rain though it hailed for some time. After the hail stopped there was no wind at all. It was a very strange quiet.

The sky was very dark but there was a very distinct edge to the storm beyond which was clear blue sky. There was one portion of the cloud that looked strange. It was spinning but it did not look like the pictures that I have seen of tornadoes. I know now that it was the rotating part of the thunderstorm from which the tornado developed. This part of the cloud spun rapidly for a while and then a gray funnel started to form. It did not look like it dropped from the cloud but rather looked like it was forming under the cloud then added itself to the cloud making the whole thing lower. It took very little time until it touched the ground about two miles to the west of where I stood.

The tornado would have hit the town of Pratt except that the funnel was located on the outer part of the rotating thunderstorm so that it's path on the ground curved from the southwest to the north. The tornado went north about five miles to hit the Pratt airport destroying a company that makes large storage tanks and two hangers. There the tornado went across the highway another two miles and hit the town of Iuka, Kansas. There it destroyed several houses, trees and a grain elevator. The tornado traveled another 23 miles before it lifted into the clouds.

One of the strange things that this tornado did was that several times the bottom half of the funnel broke off and went its own way for a while. The upper part of the storm then dropped to the ground. So for a while there were two bottoms and one top. While the tornado was in the town of Iuka, it picked up a car with a man in it who was warning people of the storm. It set the car down gently doing no damage but sucking the oil out of the car.

A friend of mine was in one of the houses that was hit. When the tornado struck the house she said that the house just exploded around her and she actually noticed the barometer on the wall fall drastically. This was probably a typical tornado, but it is one experience that I will remember for a long time."

Most tornadoes originate in the boundary between two different kinds of air masses. The location of tornado activity is farther south in the winter time than during the summer. If there are tornadoes during the summer they are more likely in the northern United States where the boundary exists between the cold and warm air masses. On a typical day in May when tornadoes occur the stage is set by the movement

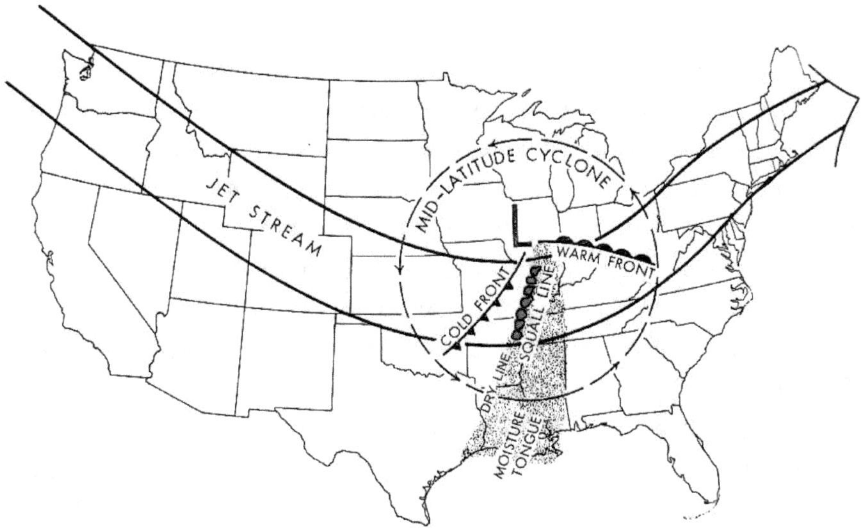

Figure 2-1. The typical atmospheric conditions that combine to allow thunderstorms to become severe and form tornadoes.

of warm-moist air up from the Gulf of Mexico (Figure 2-1). At the same time cool dry air is moving from the Northwest over the Rocky Mountains. A large low pressure center or midlatitude cyclone is also normally present. As the surface air flows around the low pressure area sharp boundaries are created between the cooler air and the warm-moist air from the south.

The jetstream is usually above the area of major tornado activity with a velocity greater than 40 mph with the stream of air blowing from west or southwest to east. Tornado activity frequently occurs where the jetstream is located above a moisture tongue. The moisture tongue is a wedge of very humid air moving from the Gulf of Mexico toward and counterclockwise around a surface low-pressure system that is moving into in the central United States.

There are at least six different factors that can be identified as very important for severe weather formation. These include a layer of warm-moist air beneath a layer of cool-dry air; a surface moisture tongue extending northward in a narrow band from the Gulf of Mexico; an eastward bound stream of higher velocity jetstream winds that is located about eight miles above the surface; proper positioning of the moisture tongue and jetstream so that they would intersect if they were at the same height; an unstable atmosphere leading to increased buoyancy and vertical motion. The final push into action may occur as the warm air in the lower atmosphere is subjected to lifting, either by an advancing cold front or dry line (line separating warm-moist air from warm-dry air from the southwest).

The normal progression of events on a day with severe thunderstorms starts with a gradual buildup of convection. As the day progresses and surface temperatures increase, small puffy cumulus clouds begin to form (Figure 2-2). These

Figure 2-2. Cumulus clouds starting to grow in a moisture rich atmosphere. (NOAA Photograph)

clouds may last for only a short period of time at first. As old ones evaporate, new ones form. As the sun provides additional heat to the ground, increasing the surface temperature, the air becomes more unstable and it is possible for clouds to last longer and grow bigger. In order for large thunderstorms to develop the atmosphere must be both unstable and moist, thus providing an additional supply of heat as water vapor condenses to liquid water inside the clouds.

A high relative humidity is important as a supply of moisture, as well as providing an environment that will allow a rising cloud to last longer without evaporation. During this development stage of severe thunderstorm activity, clouds are generally only a mile or so in diameter with vertical growth extending to less than 30,000 feet. The greatest updrafts are in the central part of the cloud and near the top of the developing cumulus cloud.

The mature stage (Figure 2-3) begins as the cloud develops larger in both diameter and vertical extent and as rain begins to fall from the cloud base. The mature stage consists of a more organized storm system. Part of the thunderstorm is converted into a rainfall generator.

Figure 2-3. Mature thunderstorms such as shown here, generate rain, lightning and may be able to generate hail and tornadoes.

Raindrops in the thunderstorm begin to grow as the air is cooled to its saturation point. Condensation on various size salt particles and dust particles in the atmosphere help to generate large drops as well as small ones. With cloud droplets of various sizes in the thunderstorm, some will be falling faster than others. This gives rise to an important growth mechanism for rain drops. Growth takes place by collision and coalescence because large drops fall faster and collect smaller cloud droplets in their path.

The area of major rainfall within the thunderstorm is located in an organized downdraft. The falling raindrops contribute to the outflow of cooler air from the base of a thunderstorm. Gusts of cold air are one of the first surface indications of a severe thunderstorm moving through a particular location. Small downdraft areas contribute to a strange pouchy appearance of the cloud that frequently indicates severe weather (Figure 2-4).

Figure 2-4. The tornado sky (mammatus clouds) frequently can be seen in clouds that form a tornado.

The mature thunderstorm that reached the severe stage is called a supercell and is a much more steady-state system than thunderstorms in other stages (Figure 2-5). The downdraft is just one part of the system. Another major part of the organized system is the updraft with associated counter-clockwise horizontal rotation. The updraft is normally located in the southwestern half of a thunderstorm moving toward the northeast. The size of the severe thunderstorm at this stage may be ten miles or more across.

Figure 2-5. The internal structure of a severe thunderstorm consists of large updrafts and downdrafts and organized circulation. View is for the storm moving toward you.

The internal structure of the thunderstorm is generated by a combination of the buoyant warm-moist air from low levels of the thunderstorm which is colliding with the stronger winds higher in the atmosphere. Winds during severe weather activity typically consist of surface wind from the south with stronger winds from the west located higher in the atmosphere as was shown in Figure 2-1. The interaction of these opposing winds helps to create an organized severe thunderstorm.

As the warm-moist air rises and begins to block the stronger winds in the upper atmosphere the winds around the sides of the thunderstorm feed clockwise circulation into it on the left and counterclockwise circulation into it on the right side looking downwind at the back of the thunderstorm as it is pushed along. An average thunderstorm moves at about 30 mph which is faster than the surface wind and slower than the winds above.

Warm-moist air is pushed into the lower part of the thunderstorm to rise and collide with the winds that are pushing the thunderstorm along as shown in Figure 2-5. Therefore, some of the warm-moist air is forced into a circular pattern as it travels around the sides of the thunderstorm. Two circulations are created within the thunderstorm with a horizontal slice through the center showing a double vortex internal structure with clockwise circulation on the right and counterclockwise circulation on the left side of a storm viewed coming toward you as in Figure 2-5. The rain area is in the right front part of the thunderstorm and contains the major precipitation with some clockwise circulation. The main updraft area is in the trailing left part of the thunderstorm and has much stronger counterclockwise circulation. This part of the thunderstorm feeds the counterclockwise rotating tornado.

The thunderstorm is able to generate a tornado because of counterclockwise circulation in the southern part of the thunderstorm as an updraft is added. This updraft arises from the winds blowing over the top of the rotating part of the thunderstorm and as the inflow of air from near the surface adds an updraft to the circulating air. The airflow over the top of the circulation creates a pull or a suction through the center of the cyclonic rotation creating a tornado in the southern part of the thunderstorm. The intensity of a tornado is related to the strength of the larger circulation within the thunderstorm as well as to the updraft within this circulation.

In the mature stage the thunderstorm may extend more than ten miles up through the atmosphere. The horizontal extent of the counterclockwise circulation within the thunderstorm may be several miles across. Persons observing a severe thunderstorm frequently describe seeing two clouds colliding. This may be due to the two circulations from different directions coming together in the leading edge of the thunderstorm as shown in Figures 2-3 and 2-5.

The final stage of the thunderstorm, the dissipating stage, begins as the updrafts weaken within the thunderstorm and a large part of the cloud is converted to downdrafts. This happens as the thunderstorm moves out of the proper environment for its continued existence so that it is not able to get enough warm-moist air into the lower front part of the thunderstorm or it does not have enough airflow around the thunderstorm to feed the circulation for its continued existence. As the cloud begins to dissolve the last flashes of lightning fade away.

It is interesting to note that a laboratory tornado-like vortex was created by the author by generating an updraft through the center of horizontal air currents (Figure 2-6). This produced a tornado-like vortex extending from the 9 ft. ceiling down to the floor of the laboratory. This verified that the specific requirements for a tornado are the creation of an updraft within opposing horizontal air currents. This laboratory tornado has been of great interest to others and has been filmed by TV stations as far away as Tokyo Japan and

Figure 2-6. Laboratory tornado generated by horizontal air currents along with an updraft above the 9 ft. ceiling.

has been used by Universal Studios and the Exploratium as will be explained in more detail later.

Severe and average thunderstorms follow the same basic pattern of development and decay except that severe thunderstorms have a more organized structure and are larger, last longer, and are much more intense.

The radar screen frequently shows a hook shape for the thunderstorm containing a tornado, to be described later, while the average thunderstorm is typically egg shaped with irregular edges. The average thunderstorm has some of the essential atmospheric components missing such as insufficient inflow of warm-moist air or wind patterns that are not quite right for tornado development.

Severe thunderstorms may form as single isolated storms or in groups of thunderstorms arranged in lines of storm cells called squall lines. A line of thunderstorms composing a squall line may extend for many miles in length. The squalls or high gusting surface winds associated with these moving thunderstorms give them the name squall lines.

Squall lines usually form a few hundred miles in advance of a cold front moving into warm-moist air. The leading edge of a squall line strikes swiftly with strong winds and rain. The weather changes rapidly but may soon return to that prior to the squall line with the trailing cold front bringing cooler weather within a few hours.

A typical cold front in the northern hemisphere extends southward or southwesterly out of the center of a low-pressure system as was shown in Figure 2-1. Squall lines forming ahead of the cold front usually run in a southwest to northeast or south to north direction parallel to the cold front. There may be several roughly parallel squall lines instead of a single well developed one. Squall lines are most frequent in the late afternoon and early evening hours. Unstable air with a high humidity content is necessary for their formation. Squall lines are common in some part of the United States from March through September, with more severe activity in the Spring than in other seasons of the year. Severe activity is more likely in the southern United States in the early spring and in the northern United States in summer. Squall lines are very typical in the Great Plains region where the supply of

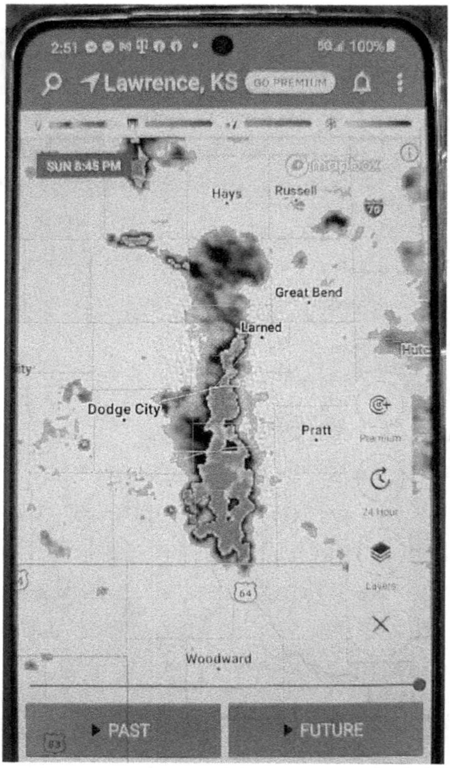

Figure 2-7. Squall lines can be viewed in real time on cell phone apps.

moisture from the Gulf of Mexico provides warm-humid air with high velocity winds flowing over the Rocky Mountains to provide energy for the storms.

Squall lines can be viewed in real time on your cell phone with apps such as that of the Weather Channel as shown in Figure 2-7. You may be able to view past and future animated movement and see if a particularly intense part of the storm is going to hit your location. Future movement is, of course, the forecasted motion while past motion has been measured.

Squall lines and large thunderstorms are regarded with extreme caution by pilots since they have severe turbulence with large updrafts and downdrafts. They can present a formidable and dangerous obstacle to flight. Less intense squall lines may have breaks between the cells that permit passage by an aircraft, but the most extensive and well developed squall lines are of such intensity that they force grounding of small aircraft. Larger airplanes equipped with radar may be able to navigate through squall lines by locating the less intense cells and the non-hail portions of the line.

Two separate areas within a non-tornadic thunderstorm may contain high velocity winds regardless of whether it is a single storm or in a squall line. The first is associated with the

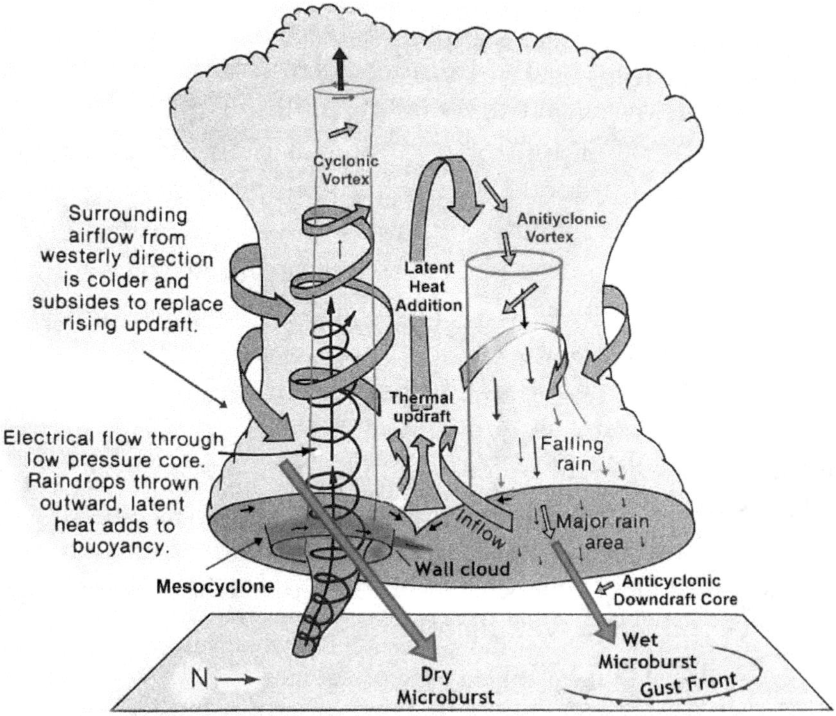

Figure 2-8. Many factors act together to form a tornado. The inflow of warm air in front of the storm encounters higher level flow from an opposite direction thereby initiating circulation. Electrical flow and latent heat contribute to the strength of the circulation after its formation.

down rush of air in the rain area of the thunderstorm that is ordinarily only a gust front but forms a microburst when it is stronger (Figure 2-8). This is the leading northeast part of a typical thunderstorm. This air has been cooled by evaporation of raindrops falling through the cloud so that a large downdraft strikes the surface and creates a gust front beneath the leading edge of the thunderstorm. These winds are frequently the first noticeable weather change at the surface due to an approaching thunderstorm. The surface winds accompanying a large severe thunderstorm may be sufficiently violent to cause considerable damage as a microburst without the presence of a tornado.

The other part of the thunderstorm which occasionally contains high wind velocities is the trailing right part of the thunderstorm. This area may contain a high velocity wind because of the counterclockwise circulation there, even when the storm is not sufficiently intense to develop a tornado. The strong circulation helps bring the surrounding colder more dense air from aloft to the surface as a microburst.

It is sometimes observed that thunderstorms without tornadoes can leave a damage path a few miles in diameter with winds as high as one hundred miles per hour. These damage strips several miles in length may accompany large thunderstorms because of strong counterclockwise circulation in the right rear part of the storm that produces a dry microburst or because of an intense downdraft in the rain area that produces a wet microburst.

The average property loss associated with severe thunderstorms is estimated to be more than $10 billion per year. These losses result from damages from tornadoes, lightning and hail in the United States. Each one of these can individually cause several billion dollars damages in a single year.

The Weather Prediction Center makes the severe thunderstorm forecast maps available every day at:

https://www.wpc.ncep.noaa.gov/national_forecast/natfcst.php

An example of this forecast map for a specific day, April 28, 2021 is shown in Figure 2-9. The regions where severe thunderstorms were likely on this day were in the warm air sector southeast from low-pressure centers over northern Texas and the Great Lakes. The threat of severe thunderstorms exists over much of Texas, Oklahoma and Pennsylvania. This map also shows the forecast for many

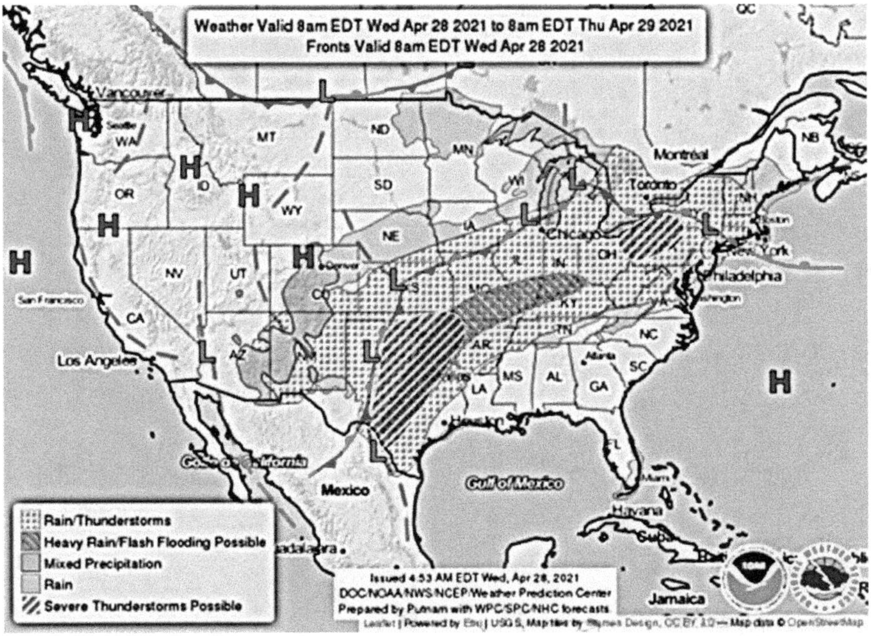

Figure 2-9. The weather forecast map for April 28, 2021 showing a severe thunderstorm forecast for Texas, Oklahoma and Pennsylvania.

other weather features for the day. On this particular day there are forecasts for regions of rain, rain with ordinary thunderstorms, mixed precipitation and flash flooding. On some days there are no areas of the United States where severe thunderstorms are forecast.

You can see a video of the winds inside a thunderstorm with tornado formation at: https://www.youtube.com/watch?v=uV-MboO_fr4&t=27s

3. WHAT YOU SHOULD KNOW ABOUT LIGHTNING

Lightning is associated with thunderstorms because of the strong updrafts and downdrafts that also produce hail and tornadoes in extreme cases. Down through history people have always had a wholesome respect for the power of the lightning flash. In fact, long ago it was believed that lightning was produced by a magical thunderbird which dived from clouds and left claw marks on whatever was struck. In the 18th century one of the big problems was that churches were constructed with very tall steeples and they were also a place that was used for storing gunpowder. Many incidences have been recorded where several people were killed and a lot of damage occurred because of this combination.

It is also interesting to know that the temple built by Solomon in Jerusalem, according to historians, experienced no damage from lightning over ten centuries. It was covered inside and out with gold and had long spikes on the dome that gave it complete protection from lightning much like the lightning rods used today.

Benjamin Franklin is credited with discovering many of the unique behavioral characteristics of lightning and electricity. He discovered positive and negative charges and studied the discharge from points which led him to develop the lightning rod.

Lightning is generated by thunderstorms that contain large updrafts and downdrafts. The charge formation is associated with ice pellets within the cloud. As ice pellets and soft hail are being formed within a thunderstorm and carried by an updraft to much colder air the outside of the pellet is colder than the inside, causing positive charges to move to the colder temperatures because of the temperature gradient. As the pellet freezes and bursts it shatters, thus giving a means of separating electrical charges in the cloud. By this process, along with induced charges on ice crystals being affected by updrafts, the thunderstorm is able to generate a significantly large charge separation within the cloud within about 20 minutes so that the base of the cloud becomes very negatively charged and the top of the cloud becomes positively charged.

The negative charges at the base of the thunderstorm are so strong that they induce a positive charge on the earth's surface even though it ordinally has a slightly negative charge. This positive charge is greater on taller objects. The lightning flash occurs when the charge separation is sufficiently great between the cloud base and ground much like a long electrical spark. A leader stroke starts down from the cloud and builds a channel of very hot ionized air. Negative charges then flow down from the cloud and positive charges from the earth flow to the cloud in a series of flashes that last less than a second. As the charge separation between the cloud and the earth builds in intensity positive charges from tall buildings and trees are drawn upward

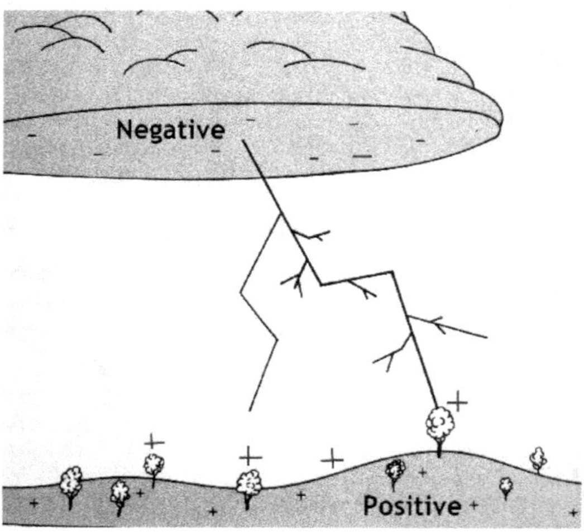

Figure 3-1. The lightning discharge develops when a large charge separation is formed between the cloud base and the earth's surface.

toward the cloud making them more likely to be struct (Figure 3-1).

About three-fourths of the energy of the lightning flash is used to heat a channel of air. The temperature within this conducting channel rises in a few ten-millionths of a second to a very hot temperature of several thousand degrees centigrade. The air expands explosively and creates sound waves that we hear as thunder.

Thunder can be heard for a distance of about seven miles and travels at the rate of about one mile per five seconds so that the distance to the lightning flash can be determined if lightning is visible since the speed of light travel is negligible. Thus, if you see the lightning and count to five and hear thunder it is one mile away. This rule may lose accuracy because some of the discharge creating the noise may come from parts of the cloud that are farther or closer than the location where the lightning strikes the ground. Rumbling thunder is created as discharges take place within various parts of the cloud.

In summary, thunder and lightning occur as a charge separation is generated within a thunderstorm that becomes quit large between the negatively charged cloud base and the positively induced charge at the earth's surface. A conductive channel of air is created as it is heated so hot that molecules of air become ionized so that electricity can be conducted very quickly. A normal lightning flash lasts for about one second and is composed of several distinctive strokes, usually three strokes, at less than one third of a second apart. These are along the same path since the air is heated by the first leader stroke providing a conducting channel of ionized air. The separate strokes may cause the flash to appear to flicker.

Several different types of lightning may be seen. One is an **air discharge** (Figure 3-2) that occurs as several leader strokes extends out from the thunderstorm and never reach the surface. This is very common in dry locations but can occur anywhere.

Figure 3-2. Air discharges occur as lightning paths do not reach the ground.

The most common lightning is **forked lightning** (Figure 3-3) that follows a zagged path from the cloud base down to the earth because the leader stroke branches out in various directions searching for the best conducting path until it finally establishes a channel down to the earth's surface. After this channel is established negative charges flow from the cloud base downward and positive charges from the earth's surface upward.

Another type of lightning is **ribbon lightning**. This is less common but occurs with winds that blow fast enough to displace the conducting channel of air between separate strokes within a single lightning flash causing a ribbon like appearance.

A **bolt from the blue** or **superbolt** is a discharge from the upper positively charged part of a thunderstorm. The ground is normally slightly negative so the bolt must extend far enough away from the storm to get away from

Figure 3-3. Forked lightning. The most common form of lightning.

the induced positive charge under the storm. After traveling outward from the thunderstorm a few miles the leader reaches the ground producing a lightning strike (Figure 3-4). The point that is struck may be as many as

Figure 3-4. Occasionally a superbolt lightning discharge occurs from the positively charged top of a thunderstorm to the negatively charged earth some distance away. (NOAA Photograph)

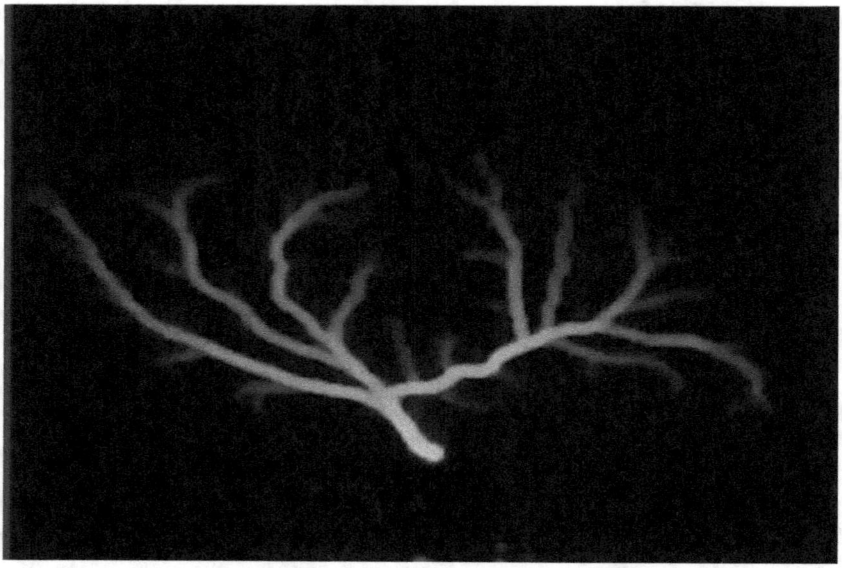

Figure 3-5. St Elmo's fire observed on NOAA's Hurricane Hunter plane.

10 miles away from the thunderstorm. The bolt from the blue is more dangerous because it is produced by a thunderstorm that is much farther away than normal and has to be stronger to travel that distance.

St. Elmo's fire is occasionally observed. This is a glow that radiates from a tall pointed object. The glow arises because of a continuous flow of positive charges from tall objects. It is sometimes observed from the mast of ships and has rarely been observed from the horns of cattle. It frequently forms on aircraft in thunderstorms as in Figure 3-5. It has been observed on the tops of thunderstorms from space.

Protective measures against lightning consist of putting lightning rods on buildings and other tall stationary objects and having the proper understanding of the nature of lightning so that shelter can be sought during thunderstorms. A lightning rod (Figure 3-6) is simply a copper or iron conducting rod that is placed on the tallest part of a building with connecting cable to the ground where it is buried deep enough to get down to the moist earth (normally two or three feet). Any lightning that would strike this building will first hit this rod and be carried down to the ground without causing damage. Sharp bends should be avoided in lightning rods as the charge is moving so fast that it may jump if there are very sharp bends.

The most dangerous places for injury or death from lightning inside buildings are near telephones, water taps, stoves, washers, and other masses of metal. In the outside dangerous places are along wire fences, under isolated trees on hill tops or on tractors with metal equipment in the ground. Much safer places are valleys, at the foot of a cliff, or under small trees in a grove of very large trees that are some distance away. There is little danger of being injured by lightning inside an automobile since the electrical charges would stay outside the automobile on the metal and go to the ground.

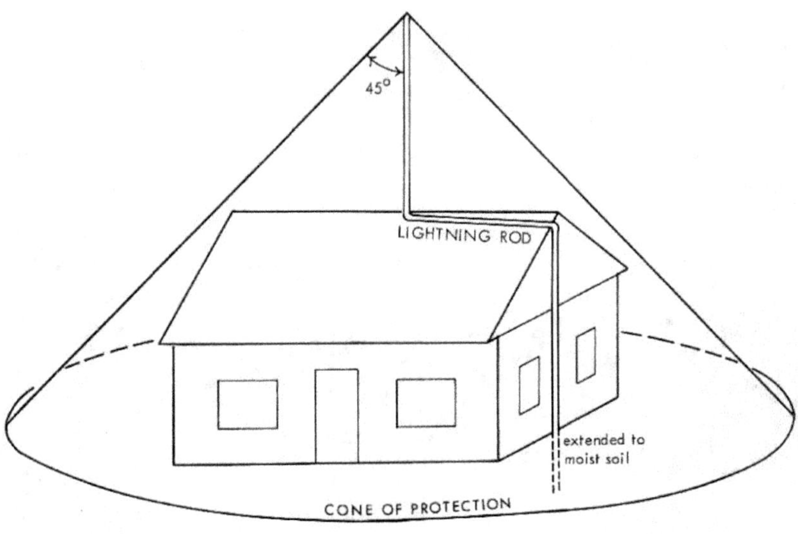

Figure 3-6. Lightning rods offer protection for 45° from the tip. Several shorter rods are usually used instead of one tall one.

The spread of electricity from a lightning strike on a golf course can be seen in Figure 3-7 from the burned grass. The flag was obviously not well grounded in deep moist soil and the electricity was neutralized as it spread horizontally. This effect would be expected if only the first few inches of soil were moist with a dry layer beneath. This spread of electricity shows the danger of a lightning strike several yards away from the point of impact.

Lightning deaths occur more frequently to persons in open country than inside buildings. Most of the fatalities, 52 %, occur in the open, 38% in houses or barns, and 10% under trees (Figure 3-8). Unsafe locations inside houses are near metal water pipes, metal appliances, and telephones with above ground lines. Dangerous locations outside are on mountaintops and hilltops (or the highest ground in the area), under isolated trees, near wire fences and on open tractors with implements in the ground.

Figure 3-7. Lightning currents are evident in this photograph taken three days after a lightning strike on a golf course in Tucson. The five-foot fiberglass flag pole was tall enough to attract lightning. (Courtesy of E. Philip Krider).

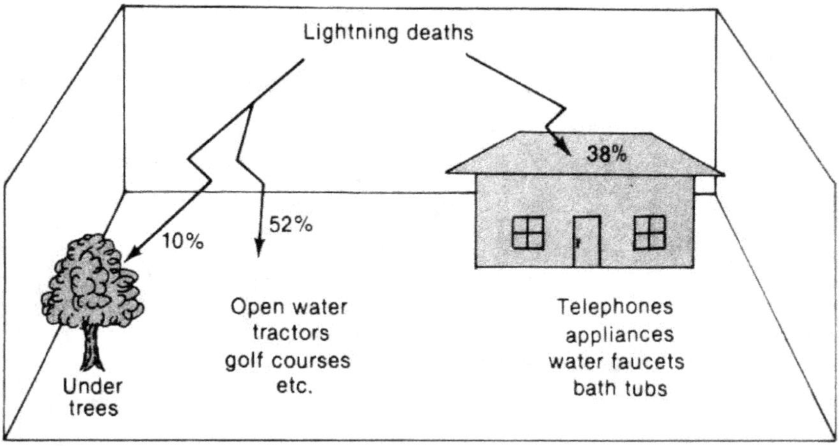

Figure 3-8. Unsafe locations during thunderstorms are under tall trees, in the open, on open tractors and golf courses, or in houses near appliances or plumbing.

Safer locations in the open are in valleys or under a small tree that is located a few hundred yards away from larger trees. Automobiles are ordinarily safe during lightning storms since they surround a person with metal. If they are struck the electricity flows through the metal to the ground but may be unsafe if lightning strikes as you are getting in or out of them.

A lightning strike does not always kill a person. In fact, only about 10% of those struck are killed but most of the other 90% are left with injuries. One farmer who was plowing during a thunderstorm was struck and knocked unconscious. He was taken to a hospital where he survived, even though the lightning had burned a hole in his hat and had also burned a streak down his back. Recovery was similar to that following a stroke.

I learned this from one of my students in my Unusual Weather class. This class filled the largest auditorium at KU holding 325 students every semester. The first time I asked if anyone had been struck by lightning, I didn't expect a positive answer. But I was wrong, because I can't recall a single semester when I didn't have one to three persons hold up their hands from out of 325 people. The hands went up timidly as would be an appropriate response to such an event. Many times, it was a friend or relative that had been struck. When I inquired about the circumstances the answers were reasonable, such as the tractor incident just described. A family member found the tractor going in circles and got the man to a hospital in time.

Lightning sometimes gives a warning a few seconds before it strikes. If your hair stands on end during a thunderstorm immediately squat as low as possible since this is an indication that you are about to be struck by lightning. Don't lay flat on the ground because the difference in potential between your head and feet may be enough to kill you if the strike is close to you as illustrated in Figure 3-7.

It is important to know that it is possible to revive many people who have been struck by lightning and are apparently dead. A lightning strike is likely to stop the heart. If there is no other damage to vital organs the victim may be revived by immediate application of cardiopulmonary resuscitation. This should be continued until an ambulance arrives.

Watch a YouTube video of lightning formation:
https://www.youtube.com/watch?v=ixoUkqKnTNs&t=130s

The following are lightning safety rules suggested by the National Center for Environmental Health:

Safety precautions outdoors

You can protect yourself from risk even if you are caught outdoors when lightning is close by.

- If the weather forecast calls for thunderstorms, postpone your outdoor trip or activity.
- Remember: **When thunder roars, go indoors**. Find a safe, enclosed shelter. Safe shelters include homes, offices, shopping centers, and hard-top vehicles with the windows rolled up.
- Don't forget the **30-30 rule.** After you see lightning, start counting to 30. If you hear thunder before you reach 30, go indoors. Suspend activities for at least 30 minutes after the last clap of thunder.
- If you are caught in an open area, act quickly to find adequate shelter. The most important action is to remove yourself from danger. Crouching or getting low to the ground can reduce your chances of being struck but does not remove you from danger.
- If you are caught outside with no safe shelter nearby, the following actions may reduce your risk:
    - Immediately get off elevated areas such as hills, mountain ridges, or peaks.
    - Never lie flat on the ground. Crouch down in a ball-like position with your head tucked and hands over your ears so that you are down low with minimal contact with the ground.
    - Never shelter under an isolated tree.
    - Never use a cliff or rocky overhang for shelter.
    - Immediately get out of and away from ponds, lakes, and other bodies of water.
    - Stay away from objects that conduct electricity (such as barbed wire fences, power lines, or windmills).
- Stay away from concrete floors or walls. Lightning can travel through any metal wires or bars in concrete walls or flooring.

Safety precautions indoors

Being indoors does not automatically protect you from lightning. In fact, about one-third of lightning-strike injuries occur indoors. Here are some tips to keep safe and reduce your risk of being struck by lightning while indoors.

- Avoid contact with water during a thunderstorm. Do NOT bathe, shower, wash dishes, or have any other contact with water during a thunderstorm. Lightning can travel through metal plumbing.
- Avoid touching electronic equipment with wired connections. Lightning can travel through electrical systems and radio and television reception systems.
- Avoid using corded phones. Corded phones are NOT safe to use during a thunderstorm. However, cordless or cellular phones are safe to use during a storm.
- Avoid concrete floors and walls. Do NOT lie on concrete floors during a thunderstorm. Also, avoid leaning on concrete walls. Lightning can travel through metal wires or bars in concrete walls or flooring.

Lightning strikes across the United States are displayed in near real time as measured by NOAA using the Geostationary Lightning Mapper from space (Figure 3-9). Time lapse coverage for the most recent two hours is displayed on the following Goes Image Viewer site: https://www.star.nesdis.noaa.gov/GOES/conus_band.php?sat=G16&band=EXTENT&length=24

On this day shown if Figure 3-9 thunderstorms with lightning are scattered throughout the central United States. The amount of lightning is related to the severity of the thunderstorms and on this day it does not indicate very

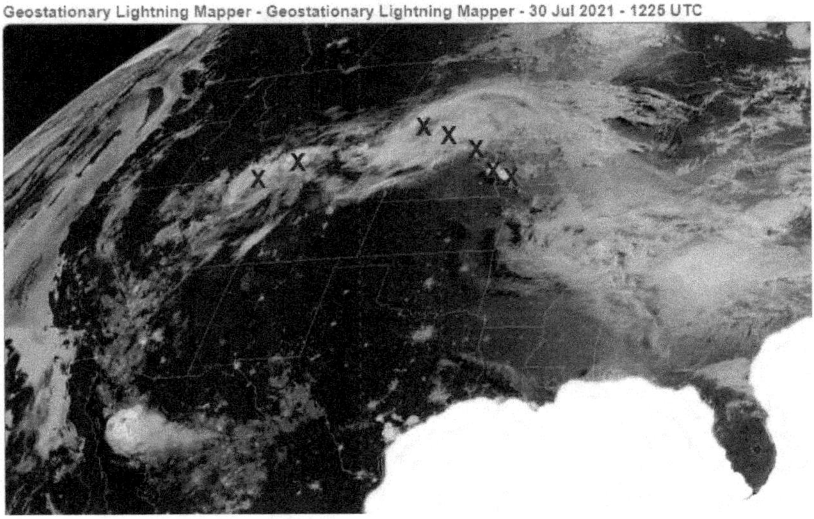

Figure 3-9. The lighting strikes are measured by the Lightning Mapper. They are shown here by X's for this day on July 30, 2021.

severe storms. Thus, measurements such as that of the lightning mapper provide additional information on the intensity of thunderstorms and weather systems.

You can watch a time lapse of lightning strikes at: https://www.youtube.com/watch?v=S-4UNSXMhDI&list=RDCMUCwF1euMcnj1jwHWzMIYBhpg&index=1

4. THE HAIL PROBLEM

Hailstones form in large thunderstorms where there are well developed updrafts and downdrafts within the thunderstorm. Since they occur with severe thunderstorms they are more likely on late afternoons on hot-humid days in the spring or summer months of the year. The same general atmospheric characteristics that develop a severe thunderstorm that contains a tornado may also produce very large hail. In fact, most thunderstorms that have tornadoes are also hail producing thunderstorms. Hailstones larger than a baseball are not uncommon. The most typical size of hail is pea size. Smaller hailstones are more likely to be spherical with larger hailstones frequently being irregular knobby shaped. The largest has been measured at eight inches in diameter and weighed almost two pounds. Even marble size hail can cause considerable damage to crops in critical growth stages while very large stones can do major damage to automobiles and even buildings, especially roofs and windows.

To be classified as **hail** it must fall from a convective storm and the particle size must be greater than or equal to 0.2 inches. The official definition of **severe hail** used by the National Weather Service is hail that is 1 inch or greater.

Figure 4-1. This photograph of the largest hailstone recorded from 1970 to 2010 shows six alternating layers made up of clear or milky appearing ice. This hailstone fell near Coffeyville, Kansas and measured over 7 1/2 in. in diameter. (Courtesy of NCAR.)

The appearance of a hailstone that is cut in half is similar to an onion, since it is composed of several layers (Figure 4-1). The layers of a hailstone consist of alternating milky opaque ice and clear ice. The environment within the thunderstorm determines the appearance of the ice within a hailstone. If the water freezes very slowly, bubbles have time to get out of the water producing a layer of clear ice. Rapid freezing traps the bubbles within the ice, producing an opaque milky layer. Rapid freezing occurs at cold temperatures high in the thunderstorm. Snowflakes and ice crystals are also likely to be present at this level. Some of these become entrapped within the outer layer as it forms on the hailstone contributing to the milky, opaque appearance of the ice formed high above

Figure 4-2 Largest hailstone ever recorded in the United States fell in the yard of Lee Scott of Vivian, South Dakota, on July 23, 2010. (NCAR Photograph)

the freezing level. Thus, the alternating rings in the hailstone correspond to the regions in the thunderstorm where the ice was formed above and below the freezing level. Small hailstones are usually round in shape, but large hailstones are seldom round and smooth. Large hailstones are often knobby and elongated. Sometimes large stones are conical shaped or flat. The most typical shape of a large hailstone is an oblate spheroid that has a very uneven and knobby outer layer.

Studies conducted in wind tunnels with ice suspended in a stream of high-speed air show some of the reasons for the different shapes of hailstones. The dynamics of airflow around chunks of ice help determine the various shapes. The optimum shape of a hailstone is related to its size and orientation within the updrafts with the dynamics of airflow dictating that a flat shape is more likely at one size and a conical shape at another. The rough outer layer of large hailstones develops as masses of water and ice accumulate and freeze because of the lower temperature of the ice as it falls from the very cold atmosphere above.

The size of hailstones ranges from 0.2″ to 8″. The record size for several decades (Figure 4-1) was the one that fell near Coffeyville, Kansas, on March 9, 1970, and measured 17″ in circumference, 7.0″ in diameter and weighed 1.7 lbs. The largest officially recognized hailstone on record to have been 'captured' in the U.S. was that which fell near Vivian, South Dakota in 2010 on July 23rd. It measured 8.0″ in diameter, 18 ½″ in circumference, and weighed in at 1.9375 pounds. Mr. Lee Scott, who collected, the monster stone originally planned to make daiquiris out of the hailstone but fortunately thought better and placed it in a freezer before turning it over to the National Weather Service for certification.

The largest recorded hailstone, prior to 1970, fell near Potter, Nebraska, on July 6, 1928 and was 5.5″ in diameter. Such hailstones are large enough to kill a person or animal. Three people have been killed by hailstones in the United States: a farmer caught in his field near Lubbock, Texas on May 13, 1930; a baby struck by large hail in Fort Collins, Colorado, on July 30, 1979; and a boater on Lake Worth, Texas, on March 29, 2000. In other countries, single hailstorms have been reported to kill more than 200 people in India and also in China.

The updrafts necessary to support hailstones of various sizes have been calculated. Tremendous updrafts are necessary to support the largest hailstones. A hailstone of 5″ in diameter requires an extreme updraft of 234 mph to support it (Figure 4-3).

A 5″ hailstone would not have to be completely supported by an updraft since the last ring could grow

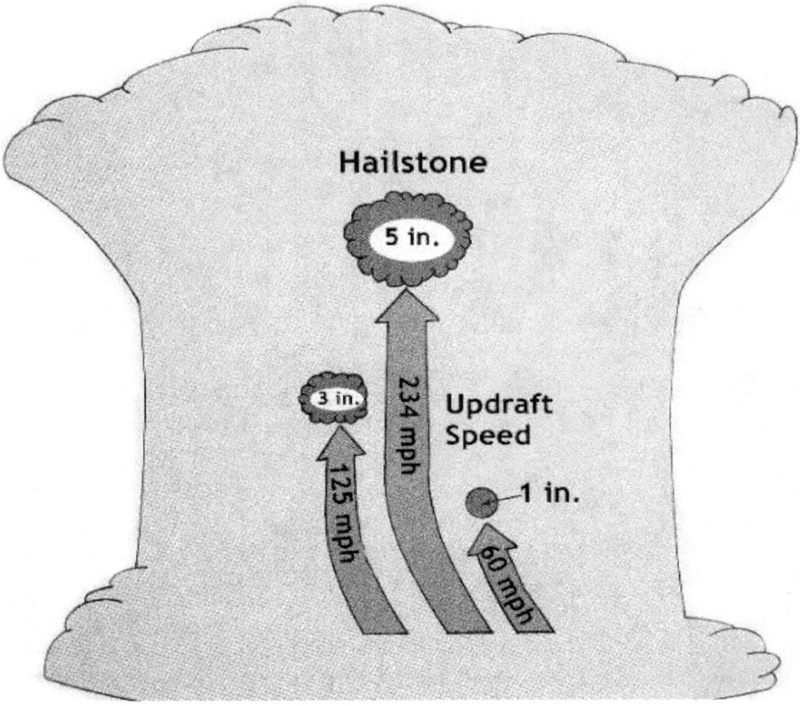

Figure 4-3. Strong updrafts are required in thunderstorms to support large hailstones.

as it fell from the cold air above the freezing level to the surface. However, the hailstone would have to be supported by an updraft of sufficient magnitude to carry it upward when its size corresponded to the next to last layer within the hailstone. Even the next to last layer of larger hailstones includes a very large mass of ice that would require extremely large updrafts within the thunderstorm of more than 100 mph.

The frequent occurrence of hailstorms shows that large updrafts are common in thunderstorms. If the upper atmospheric temperature is cold, surface heating of the lower atmosphere causes the atmospheric stability to decrease, creating an environment for the development of

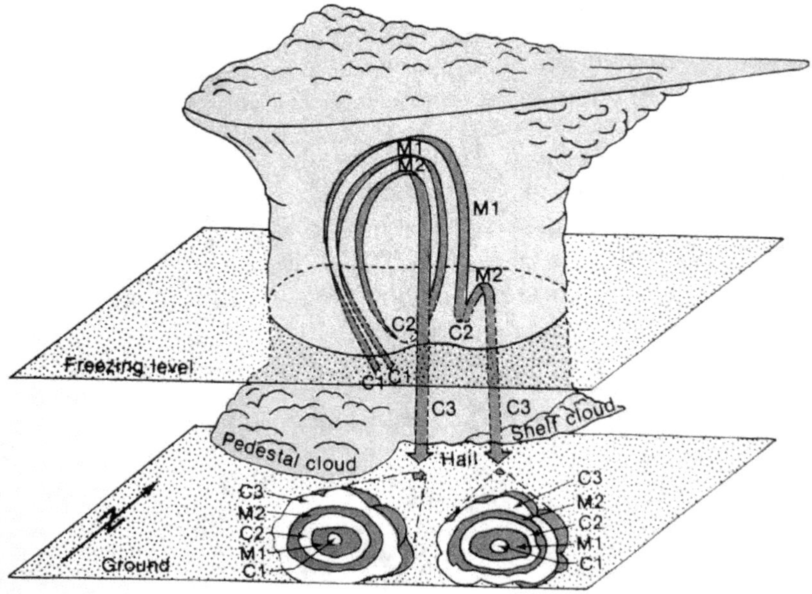

Figure 4-4. The various rings of a hailstone are formed as it circulates within a thunderstorm. Clear layers are formed where the ice freezes very slowly; milky layers are formed under more rapid freezing conditions. The number of trips through a thunderstorm determines how many rings of ice the hailstone will have.

large thunderstorms with intense updrafts. The large updrafts and downdrafts are necessary to form the different layers in a hailstone as it is circulated above and below the freezing level as shown in Figure 4-4.

Hailstorms cause considerable damage to agricultural crops each year in addition to damage to personal property and animal life. Hailstorms are the greatest problem in the central Great Plains reaching a maximum in Northeastern Colorado and Southeastern Wyoming. Hailstorms are most frequent in May and June in the late afternoon hours.

Some important atmospheric conditions that contribute to the development of hailstones are surface streamline confluence ahead of a cold front, high wind speeds within the jet stream, large wind shear and an unstable atmosphere.

Squall line thunderstorms account for the greatest percentage of the total amount of hail, but the hail from a supercell thunderstorm may be larger and more damaging. Hail is most likely to fall beneath the central part of a thunderstorm while heavy rain generally occurs beneath the leading edge of the thunderstorm, and tornadoes, if present, are more likely on the southern edge of a thunderstorm moving toward the northeast. Most hailstorms come from the southwest and travel at a speed slightly slower than the average wind throughout the atmosphere.

The worst year for hail damage was 2017 when hailstorms caused $22 billion damages. A billion-dollar single hailstorm occurred on April 10, 2001 as it traveled across central Missouri from Kansas City to St. Louis. It formed along a stationary front beneath a strong jetstream from the southwest. This supercell produced hail larger than baseballs and nine tornadoes that killed one person and damaged many houses, airplanes and automobiles.

The National Weather Service began a new severe thunderstorm warning service in 2021 that will automatically activate an alert system on smartphones within a warned area. There are three categories of severe thunderstorms in order of lowest to highest damage threat. Each tag and additional messaging is designed to promote immediate action, based on threats, according to the National Weather Service.

- **Base** is the lowest threat, with one inch or quarter sized hail and/or 58 mph thunderstorm winds. This will not activate Wireless Emergency Alerts (WEA), which means damage is expected to be at the base level.

- **Considerable** damage threat is when hail is at least golf-ball sized (1.75 inch), and/or 70 mph thunderstorm winds. This also will not activate a WEA.
- **Destructive** is the highest damage threat, with baseball sized hail (2.75 inch), and/or 80 mph thunderstorm winds. This tag will automatically activate a WEA on smartphones within the warned area.

The use of these categories by the National Weather Service should better convey the severity and potential impacts from thunderstorm winds.

You can watch hail formation in a thunderstorm on YouTube:
https://youtu.be/pSd-b42Ss6Q

5. UNDERSTANDING TORNADOES

Even the word tornado has a certain effect on people. It causes fear in some or may bring back memories in others. Those who have been through a tornado or seen the damaging effects it can have are aware of the powerful and destructive forces that they are able to generate. The wind speed within tornadoes varies with the size and intensity of the tornado from a speed less than 100 mph to an intensity that may exceed 300 mph. Tornadoes also come in various sizes. The smallest ones are only a few yards wide while the largest tornadoes are over one mile in diameter. An example of a very large damaging tornado is the one that occurred in Topeka, Kansas, and was shown in Figure 1-1. This storm has the distinction of being one of the first resulting in over $100 million damages. Damages adjusted to today's dollars are $1.8 billion making it rank as one of the costliest single tornadoes. The wind speeds were estimated to be over 250 mph.

The typical tornado stays on the ground for only about ten miles with the longest ones staying on the ground for over two hundred miles. The most severe tornado from the standpoint of human injuries and death was one of these long path tornadoes that was on the ground for 219 miles on March 18, 1925. This tornado traveled through Missouri, Illinois and Indiana, killing 695 people and injuring many thousands.

Today, people are better able to protect themselves from tornadoes than in 1925 because of tornado forecasts and real time availability of weather information. Only 17 people died from the Topeka tornado and one of these was from a heart attack. This was in spite of more than a thousand houses being completely destroyed.

The loud noise produced by tornadoes has already been described as a warning of an approaching tornado. The noise is usually described as very loud; like hundreds of jets or freight trains. The noise probably comes from the high-speed winds rather than interaction of the tornado with buildings or whatever else is in the way.

Tornadoes occur in the southwest part of a thunderstorm which is typically moving toward the northeast with the major rain area in the leading northeast part of the thunderstorm. The appearance of the tornado funnel depends somewhat on the direction of viewing, the backdrop behind it, and the material that is picked up from the surface. If the whole thunderstorm, that may be more than ten miles in diameter, is to the northeast of the observer the tornado viewed against the dark rain area in the background appears white (Figure 5-1, top right). If the major part of the thunderstorm is to the west of the observer, the tornado normally appears darker against the lighter background behind the thunderstorm.

Tornadoes may take on different colors as they pick up material from the surface. For example, if they are moving across red colored soil they may pick up a lot of red dust and the whole funnel takes on a red hue. Typically the debris and darker soil causes them to have a darker color than the surrounding background.

A tornado is visible because of condensed water vapor, in addition to the dust and debris picked up from the ground. Water vapor inside the funnel condenses because the pressure is less. The lower pressure inside the funnel causes the air to expand and cool with the result that water droplets condense. Figure 5-1 shows some of the different types of funnels. Those on the left have debris or dust in the lower part of the funnel and condensed water vapor in the upper part to make them visible. The funnel may vary from massive as in the middle photographs to rope shape as in the top photographs.

Figure 5-1 Some of the different shapes and sizes of tornadoes. (Composite of NOAA photographs).

Sometimes tornadoes glow from electricity. It has been observed that thunderstorms that produce tornadoes also have very high electrical activity associated with them. The very large severe thunderstorms have higher velocity updrafts and downdrafts that are very efficient in producing hail and lightning as well as tornadoes.

The shape of a tornado can vary from an almost completely cylindrical shape to a very extreme funnel shape. The typical tornado has a larger diameter near the cloud base that narrows down to a smaller diameter by the time it gets to the ground as shown by most of those in Figure 5-1. A tornado vortex that does not extend all the way to the ground is called a funnel cloud. A thunderstorm that is barely able to generate a tornado may develop a tornado that is not able to stay on the ground and skips along the surface because of insufficient strength in the thunderstorm to continue to support a tornado. The last stage before dissipation of the tornado generally has a rope type appearance.

Tornadoes are assigned a category by using the Enhanced Fujita scale to estimate it's strength. It replaced the Fujita scale (F0-F5) and became operational on February 1, 2007. It is used to assign each tornado a rating based on the estimated wind speeds determined by damage. When tornado damage is surveyed the amount of damage to various types of buildings and objects is used to assign a rating from EF0 to EF5. Estimates of the ability of different types of structures to withstand winds of a particular speed allow wind speed estimates to be assigned to each category.

EF-Scale	Intensity	Wind mph	Damage
EF0	Light	65-85	Branches broke
EF1	Moderate	86-110	Roof damage
EF2	Significant	111-135	Roofs off
EF3	Severe	136-165	Walls torn off
EF4	Devastating	166-200	Houses leveled
EF5	Incredible	Over 200	Trees debarked

Sometimes more than one tornado forms from a single thunderstorm. The first tornado photographed was in 1884 and it had one main tornado and two smaller funnel clouds (Figure 5-2). As many as six have been observed extending from the same cloud.

There are many myths associated with tornadoes. The myth that was commonly accepted before the tornado in Topeka, Kansas in 1966 has already been mentioned. It was an Indian legend that Topeka would never be struck by a tornado because it was protected by a hill called Burnett's Mound on the southwestern edge of the city. Another myth that has been proved untrue is that tornadoes will not cross a river. An example is the tornado that struck Cincinnati, Ohio on April 3, 1974. A large tornado crossed the river from Indiana into Kentucky then crossed the river again into Ohio to strike the city of Cincinnati.

One of the reasons that such myths can persist for many years is that the probability of tornadoes striking any one spot is small. If the probability is calculated for a tornado striking

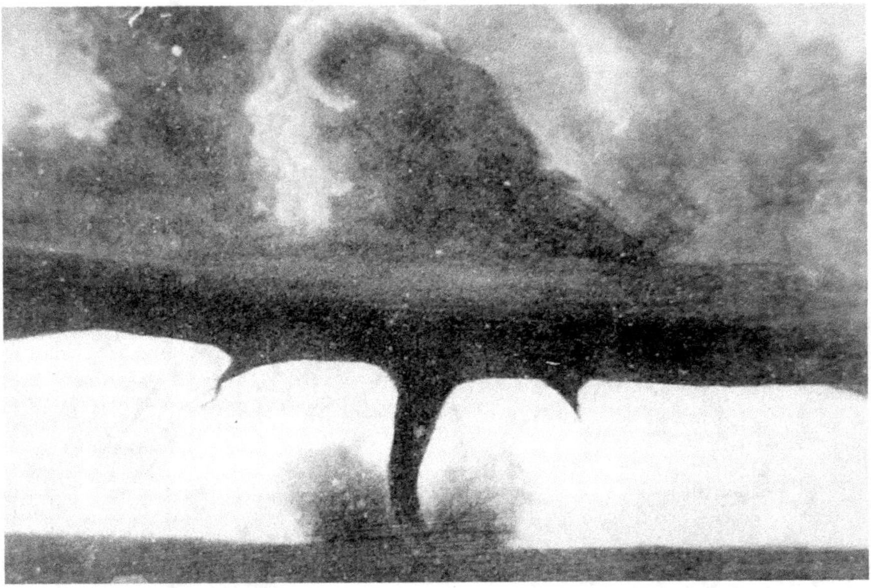

Figure 5-2 The first tornado photograph was taken in 1884 near Garnett, Kansas (NOAA).

a particular spot, even in the central United States where tornadoes are most frequent, it is only one tornado in about 500 years. This means that it takes a long time to disprove any given myth.

In spite of the low probability of a tornado striking a particular spot, tornadoes are not evenly distributed in time and they do strike some locations much more often than this. I talked to a man in Lubbock, Texas, after the tornado in 1970, who was picking up bricks from his house that had been demolished by the tornado. He told me that it was the third time that his house had been destroyed by a tornado and it was the second time on that particular spot.

The geographical distribution of tornadoes in the United States is such that the central Great Plains area has a much higher frequency than other locations (Figure 5-3). Every one of the fifty states has at least an occasional tornado. The western and Rocky Mountain states as well as Alaska and

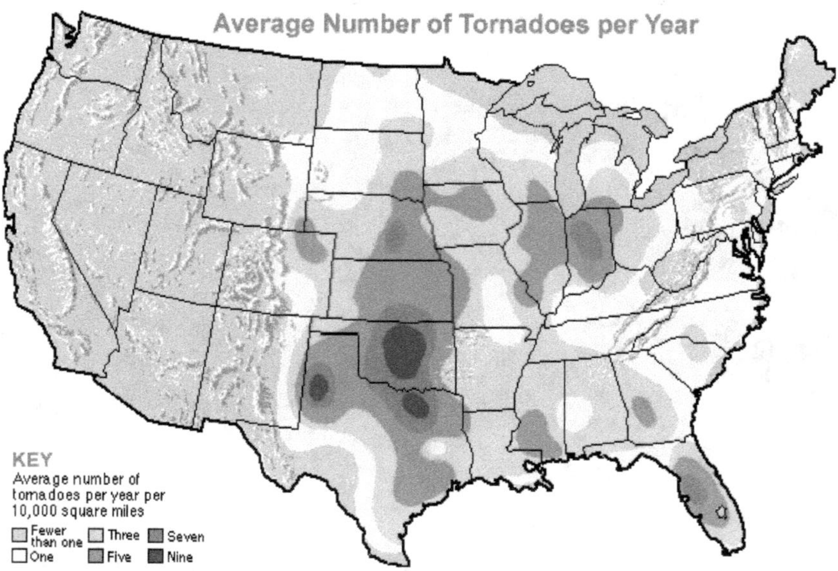

Figure 5-3. Distribution of tornadoes per square mile in the United States.(UCAR)

Hawaii have a much smaller number of tornadoes in comparison to the others. Tornadoes are most frequent in a band including Texas, Oklahoma, Kansas, Missouri, Iowa, Illinois and Indiana. This concentration of tornado activity is related to the proper combination of atmospheric variables that are necessary to cause severe weather development as previously described. The Rocky Mountains to the west allow a thin layer of warm-moist air near the surface to move northward in response to surface pressure patterns while cooler dry air is flowing over the mountains from the west. Severe weather activity is triggered in local areas as a cold front or dry line moves into an area or as solar heating causes the atmosphere to become more unstable. Florida also has a lot of tornadoes but they are not generally as strong as those in the central United States and are more related to the high humidity and surface heating.

The geographical location of maximum tornado activity varies with season. It follows the jetstream that in general separates the cooler air to the north from the warmer air masses to the south of the jetstream. If there are tornadoes in the winter they are more likely to occur in the southern United States along the gulf coast where contrasts between cold and warm air exist. In the spring and fall the area of maximum activity shifts to the central United States with a higher frequency in the spring months because of the greater intensity of the upper-level winds in the atmosphere. If tornadoes occur in the mid-summer months they are more likely in the northern United States where the greatest contrast in air masses exists and where the jetstream is more likely to be located.

Tornadoes have occurred in every month in the year, but there is a distinct peak in tornado activity during the spring months with May having a higher frequency than other months (Figure 5-4).They are least frequent in January with much less activity during this month. More days in June have tornadoes than any other month.

Since severe thunderstorm activity is tied to solar heating of the ground as one of the ways of stirring the atmosphere into major activity, tornadoes are much more likely in the late afternoon hours than any other time of the day. This is especially true in the central United States where tornadoes are most likely.

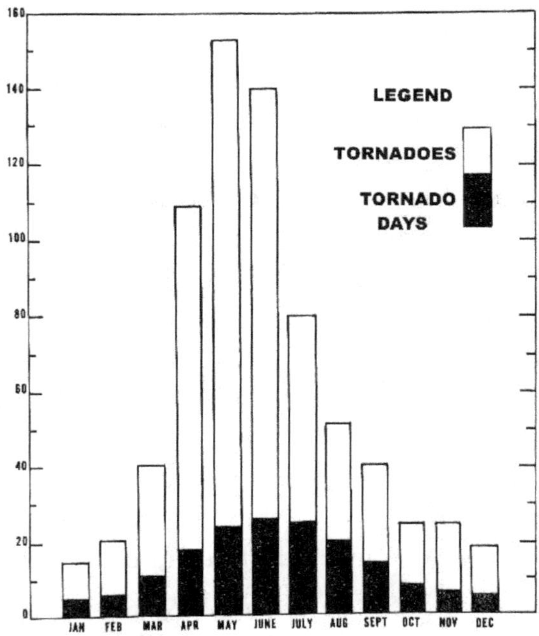

Figure 5-4. The number of tornadoes and days with tornadoes for each month.

The time of day when tornadoes are most frequent is from four to six in the afternoon. In the Southern United States near the gulf coast there is still a peak in tornado activity in the late afternoon hours although they can occur at any time during the day because of the hot-humid air that is always present.

In connection with the generation of tornadoes by thunderstorms it is of interest to learn that a tornado-like vortex can be formed in the laboratory using the same wind currents. Such a tornado was generated in my laboratory as shown previously in Figure 2-6. This was done by using air currents in the horizontal plain while applying a suction produced by an updraft through the center. The counter clockwise rotation in the southwestern part of the thunderstorm combined with an updraft through the center provides the mechanism for generation of a tornado in the

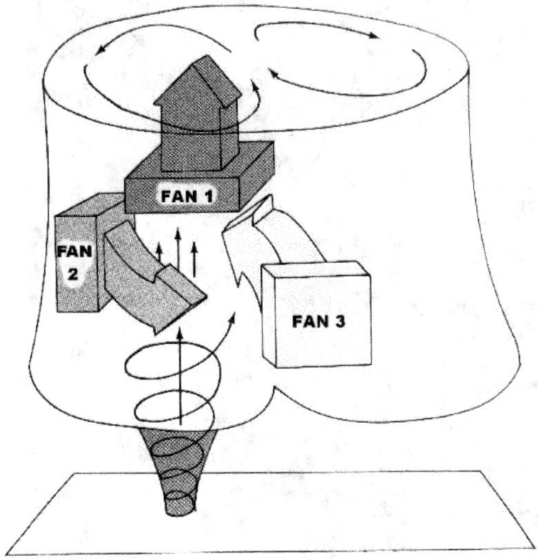

Figure 5-5. The laboratory tornado is generated by simulating the updraft within a thunderstorm and the air currents for a thunderstorm with a tornado in the southwestern part.

thunderstorm. Comparison of the formation of the tornado in a thunderstorm and in the laboratory is shown in Figure 5-5.

The laboratory tornado received a lot of attention from newspaper reporters in Lawrence, Topeka and Kansas City. It was also filmed by television stations in Kansas City, Topeka and Tokyo, Japan. It was covered in an article in the Preprint Volume of the 1971 National Conference on Severe Storms and presented to the audience at the conference. By one or more of these means it came to the attention of Ned Kahn at the Exploratorium in San Francisco in the early 1980's. He called me and asked if I could help him create a similar but larger tornado-like display. I agreed and gave him one of my books that described how I had created it in my laboratory. Over the next several months he called regularly as he struggled with the details of the display until it finally was completed. He was thrilled with the appeal of the new attraction and my book, *Severe and Unusual Weather*, that contained a chapter on creating laboratory tornadoes was sold for many years in the gift shop.

The tornado attraction has survived several transformations and continues to be of interest to many people. It is still today one of the major attractions for the Exploratorium (Figure 5-6).

Figure 5-6. The tornado display at the Exploratorium continues to attract attention from all ages: Photo credit: Esther Kutnick, Exploratorium, www.exploratorium.edu

After a movie made by Universal Studios in 1996 called Twister, they used the 9 foot laboratory model as described in books I had written to generate a 50 ft. tornado simulation, shown in Figure 5-7 and 5-8 in Orlando after settling a lawsuit that took me to Hollywood.

"Name your price," said the lawyer on the other end of the line. It was the first time that I had ever been told that. A week earlier someone from Universal Studios in Hollywood had asked me to come there for a trial. Universal Studios was being sued by Volvo Corporation over the rights to create an artificial tornado. Volvo claimed the right to produce artificial tornadoes because they had recently used one in a television commercial. I was not given all the details at first but I learned later that I was being placed in the middle of this case because I had previously written a technical book and two textbooks that described how I had produced the first unconfined tornado-like vortex by using a system of fans. The fans were placed above a hole in the ceiling of my laboratory to pull the air upward and others were attached to the ceiling to create cross winds similar to measurements within thunderstorms from Doppler radar. The resulting miniature tornado extended from the floor into the hole in the ceiling and could be seen by using steam or other means for visualization.

I realized that the tornado I had created in my laboratory at the University of Kansas was original and could probably be patented. Since I was a Professor of Atmospheric Science and had developed the tornado on University property a patent would be shared by the University. I chose to make the information on developing the tornado available to the public by publishing it in research articles and in my books.

As a Professor at the University in 1998 I was involved in the usual full-time load of teaching and research so it was a little hard for me to relate to Universal Studios urgency for

Figure 5-7. The Twister building housing the 50-foot tornado at Universal Studios in Orlando.

me to come to Hollywood so I had turned down their request. But I realized that being told, "Name your price," was different and I quickly ran through some numbers in my head and came up with $1000 per day for my services. I later realized when I became aware of more of the details that I could have asked for 10 times that amount and they would have gladly paid it.

My wife, Doris and I were given tickets to Los Angeles and accommodations in the Westin Hotel. We were in Hollywood for three days and needless to say Universal Studios won their case when one of my books, *Severe and Unusual Weather*, was entered as evidence since it contained detailed information on the creation of the artificial tornado and was first published in 1983.

After the successful conclusion of the lawsuit they were able to create a realistic looking tornado for their attraction. It was the center of attention in the Twister building from 1998 to 2015 (Figure 5-8) and was viewed by millions of people. Because it was patterned from my model, I was able to ask for and receive 18 free tickets to Universal Studios during the summer of 2010 for my wife, three children, their spouses and our ten grandchildren. The public relations person sent my request to their lawyers who still had a record of Universal Studios paying my way to Los Angeles to serve as an expert witness to defend their right to develop such a tornado simulation. This verified that my book, *Severe and Unusual Weather*, was used to show how it was developed and that it was in the public domain.

Watch a YouTube video of tornado development and see the simulated tornado in action in my laboratory: https://www.youtube.com/watch?v=LBzhAYuDnm8

You can see a YouTube video of the tornado at the Exploratorium: https://www.exploratorium.edu/video/explainers-tornado

Figure 5-8. The tornado attraction at Universal Studios being viewed by a few of the millions of visitors who saw it.

Watch a YouTube video of the 50-foot tornado in the Twister building at Universal Studios.
https://www.youtube.com/watch?v=DxpKu6Z_-NQ

Watch a computer simulation of a tornado.
https://www.youtube.com/watch?v=e2wbn3ivHwc&t=270s

6. TORNADO DEVELOPMENT AND WATCHES

The largest thunderstorms are most likely to be severe, i.e., contain frequent lightning, and damaging winds or hail. Such large thunderstorms develop in the atmosphere as localized areas of air become unstable from several factors, such as heating of the ground. As a large bubble of air becomes less dense than the surrounding air it is buoyant and rises, perhaps high enough to form a cloud. If other appropriate atmospheric conditions exist, as described previously in Chapter 2, a severe thunderstorm is produced.

The exact location where a severe thunderstorm will develop cannot be forecast since it is dependent upon such factors as the ground beneath it, airflow over a hill that initiates an updraft, or converging surface winds that initiate updrafts. Since the location where a specific thunderstorm will develop cannot be forecast, we use the next best approach consisting of forecasting the location where the appropriate atmospheric conditions exist for generating severe thunderstorms.

Severe thunderstorms develop when the proper atmospheric conditions are present. A good example is one of the worst tornado outbreaks in history. Figure 6-1 shows the weather conditions on April 3, 1974 when 148 tornadoes touched down. Thirty of them were F4 or F5

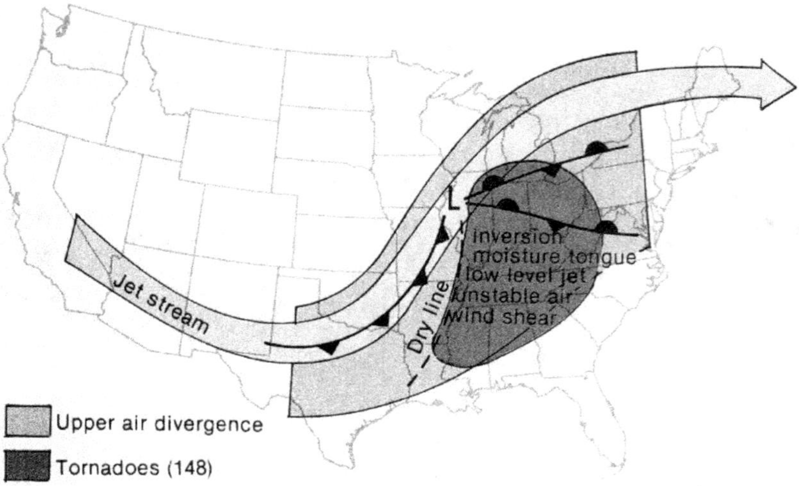

Figure 6-1. The location of tornado watches is usually in the warm air section of a frontal cyclone in advance of a dry line or cold front as in this extreme example of severe weather. They are not generally as large as in this extreme example.

comparable to today's EF4 or EF5 in strength and some of these hit cities including Xenia, Ohio. Wikipedia calls this the 1974 Super Outbreak of tornadoes with good reason as it killed 319 people. This Super Outbreak was the first tornado outbreak in recorded history to produce more than 100 tornadoes in under a 24-hour period, a feat that was not repeated until another Super Outbreak in 2011 produced 175 tornadoes but fewer strong EF4/EF5 tornadoes.

A severe thunderstorm or **tornado watch** is issued by the Storm Prediction Center located in Norman, Oklahoma, on the basis of a combination of all the appropriate atmospheric conditions previously described and shown in Figure 6-1. The intersection of the moisture tongue with a line projected downward from the axis of the jetstream is a common location of severe weather. This is typically located within the warm air sector of a frontal cyclone, and also corresponds to a region of upper level divergence. The atmosphere is frequently more unstable farther southward in the warmer air. However, severe thunderstorm activity is most likely where the most divergence exists above the moisture tongue. With these conditions, squall lines develop in advance of the cold front along the dry line.

The area described in a severe thunderstorm forecast is frequently a rectangle as shown in Figure 6-2. It is located in the warm air sector of a midlatitude cyclone and may be bounded to the north by the warm front and to the west by the cold front. Since the whole weather system is moving, the line of most likely activity continually moves toward the east; therefore, a rectangle is needed to describe the forecast area of severe weather for more than an instant in time. You can frequently tell when the danger of severe thunderstorms is past before the watch is officially ended by noting the passing of the cold front. If the wind changes direction from the southwest to the

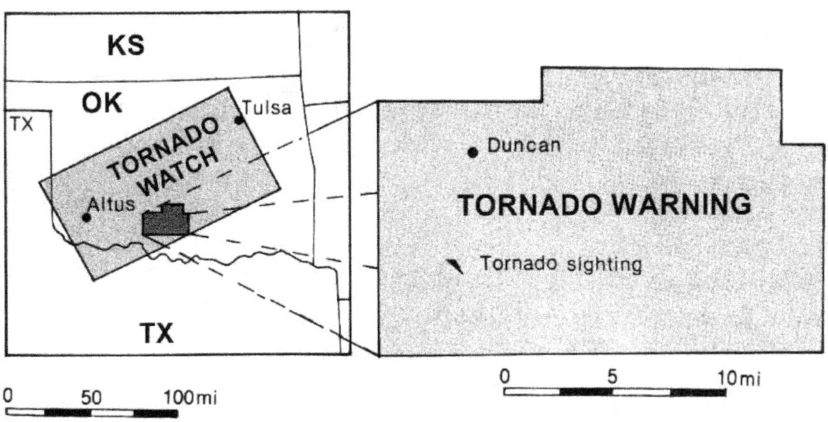

Figure 6-2. Tornado watch areas are issued on the basis of the presence of appropriate atmospheric conditions for tornado formation; thus, they cover a large geographical area. Tornado warnings are issued after a tornado is formed.

northwest, and the temperature decreases, this indicates the front has passed, with a decreased likelihood of severe weather.

Additional "inside information" is available after squall lines have developed within a severe thunderstorm watch area, since the whole watch area does not normally have thunderstorms. Existing thunderstorms can be tracked by radar and are likely to move toward the northeast because of the upper level winds. Therefore, the area most likely to have severe weather is pinpointed much more closely than the large rectangle describing the entire severe thunderstorm watch area. Once a thunderstorm develops it can be tracked on a cell phone with future movement projected from its past movement.

In some cases, a severe thunderstorm watch is issued and no storms develop. Additional triggering mechanisms such as topographic features, and uneven heating at the earth's surface are sometimes needed to start thunderstorm development.

The impact of the ordinary severe thunderstorm watch is weakened because of the size of the area. Quite often, some of the watch area does not even have cloud cover, and only a few thunderstorms within the area develop hail, high winds, or a tornado. This makes the forecast seem much worse than it really is to the many people who do not see a cloud. This effect is not as great as in the past because of the real time radar coverage of the whole area available on cell phones today.

The atmospheric conditions required for severe thunderstorms vary somewhat with geographical location. In the southern United States, the atmosphere becomes so humid and unstable that severe thunderstorms can grow in the absence of some of the other specific atmospheric conditions, such as a strong jetstream, that are otherwise required for their formation.

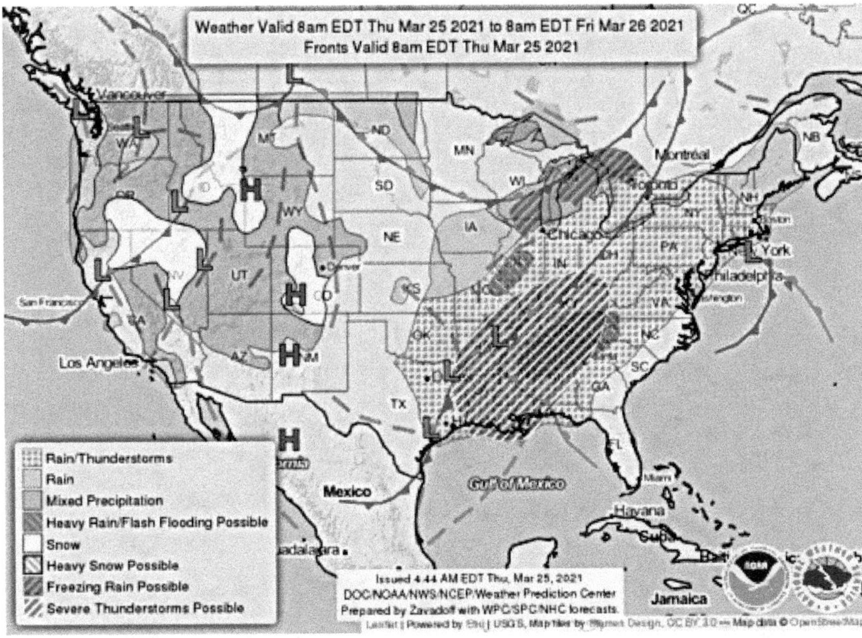

Figure 6-3. The forecast for March 25, 2021 was for the possibility of severe thunderstorms over several southern states southeast of low pressure systems.

Thus, the relative magnitude of the various atmospheric conditions is important, which makes the problem much more complex than it would be if we were dealing only with the presence or absence of all the various atmospheric conditions.

The forecast area for severe thunderstorms on a particular spring day on March 25, 2021 is shown in Figure 6-3. The possibility of severe thunderstorms was forecast for a large area of southern states. It is important to note that this map that is routinely produced early in the day shows the possibility of severe weather during the day but actual tornado watch boxes for smaller areas were forecast for shorter time periods as the day progressed.

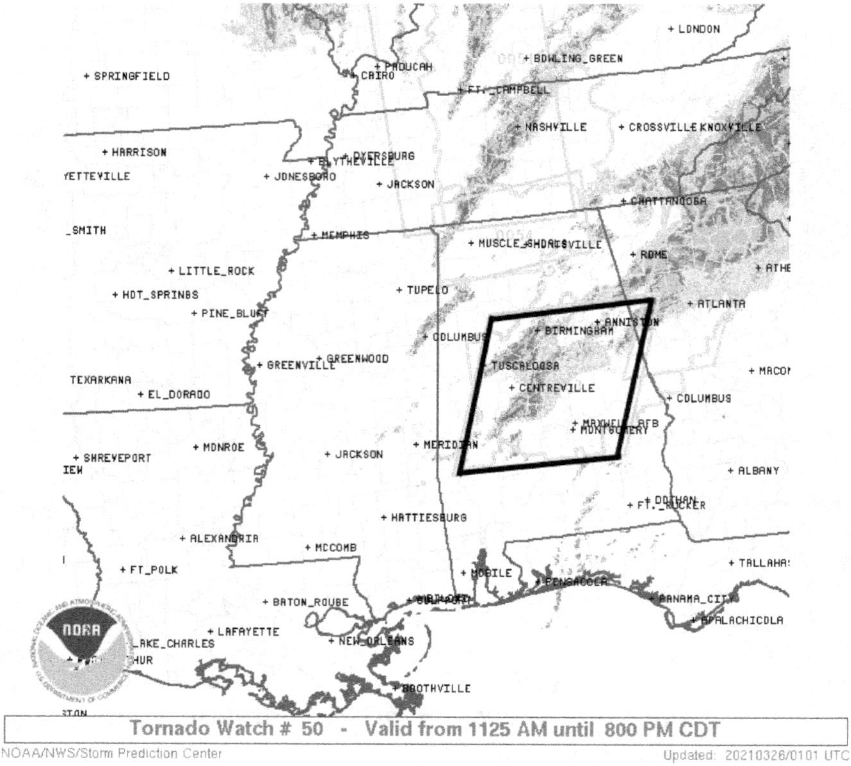

Figure 6-4. One of several tornado boxes issued on March 25, 2021. This box includes two very strong thunderstorms.

Several tornado watch boxes were issued for this day. One of them is shown in Figure 6-4. This box was issued for the two very strong thunderstorms that are in this box. One of them produced a EF2 long-track tornado that scarred the landscape across five counties during the afternoon of March 25, 2021.

There is another source of tornado watches that should be mentioned. When a hurricane or even a tropical storm approached the Gulf States it brings with it the possibility of tornadoes. The strong winds in the right front quadrant of the storm bring very warm-moist from the ocean on shore. The updraft of this air combined with the very large horizontal cross currents of air associated with

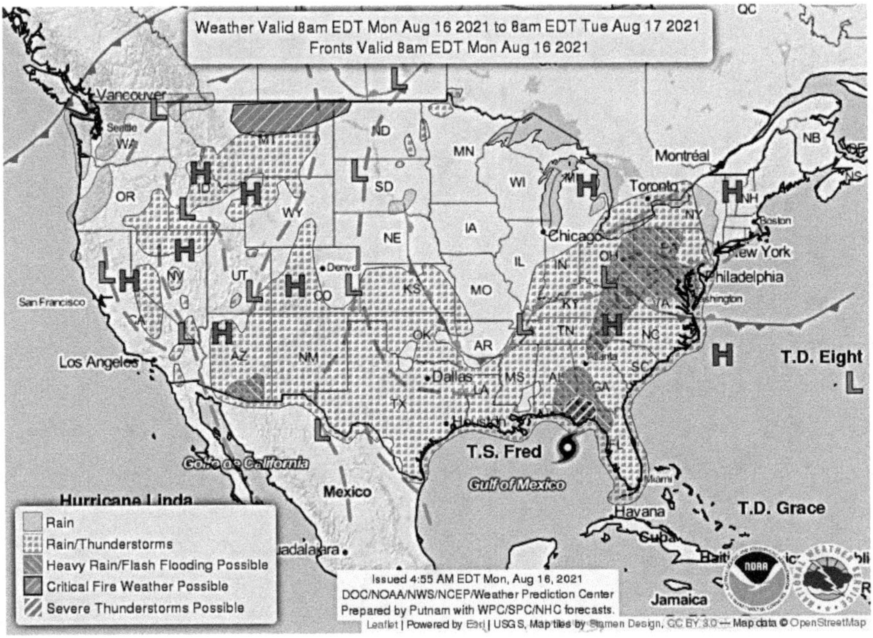

Figure 6-5. The forecast includes the threat of tornadoes in the Florida Panhandle and surrounding states as tropical storm Fred approaches land on August 16, 2021.

the tropical storm are the right ingredients to produce tornadoes. These form in thunderstorms within the rain bands of the hurricane or tropical storm. Figure 6-5 shows the forecast for severe thunderstorms associated with tropical storm Fred as it approached land on August 16, 2021. More than a dozen tornadoes touched down as the storm moved across Florida and northward through surrounding states. Several hurricanes in previous years have been accompanied by more than a hundred tornadoes.

The **tornado warning** is the subject of the next chapter and is issued when a tornado is spotted or when a distinguishing feature such as a hook echo is seen on the radar screen.

7. TORNADO DETECTION AND WARNINGS

A **tornado warning** indicates that a direct threat of tornadoes exists in a localized area. Specific warnings are relayed by the local weather service, Civil Defense Director, and television or radio station operators. They are issued on the basis of the observation of a hook echo (Figure 7-1) or other signatures on the radar screen or from eyewitness reports, normally verified by highway patrolmen. If the local weather station is equipped with radar, the formation of a hook shaped echo is the usual basis for issuing a tornado warning. The hook shape on the screen only indicates that the thunderstorm has generated a mesocyclone (cyclonic circulation) which may be capable of producing a tornado. More than half of the tornadoes occur from thunderstorms that do not produce a hook shaped echo on the radar screen. In order to see the hook, the radar must be aimed at the lower one-fourth of the thunderstorm. Thus, the range of the radar is limited for severe thunderstorm detection because of the earth's curvature.

The hook echo forms as the radar displays the reflection from raindrops within a horizontal slice through the lower part of a severe thunderstorm. The radar signal is reflected back to the station by raindrops carried in the

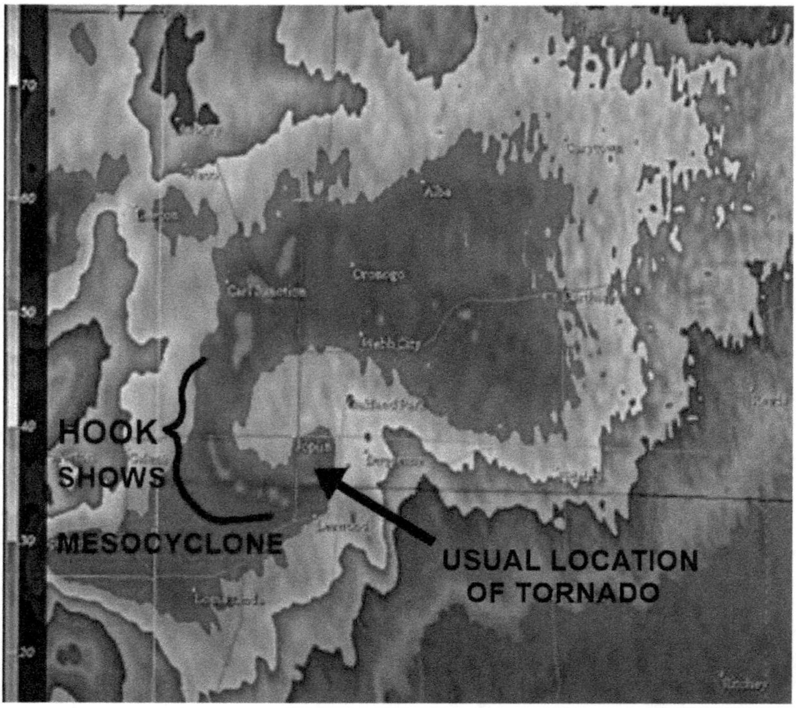

Figure 7-1. The hook echo for the devastating Joplin tornado May 22, 2011. It killed 158 people and caused damages of $2.8 billion. (NOAA Photograph)

outer part of the mesocyclone. This circulation brings rain originating in the thermal updraft around to the front part of the thunderstorm forming the hook shape on the radar screen. Most thunderstorms that form a hook echo also develop a tornado, although the tornado does not show up on a radar screen because it is much smaller than the mesocyclone that is about half the size of the thunderstorm. But a tornado may pick up enough debris to show up as a small circle such as that at the tip of the arrow in Figure 7-1.

Another feature that is useful for identifying supercells that are likely to contain a tornado is the v-notch or sometimes called the flying eagle. An example is shown in Figure 7-2. The notch that sometimes shows up on

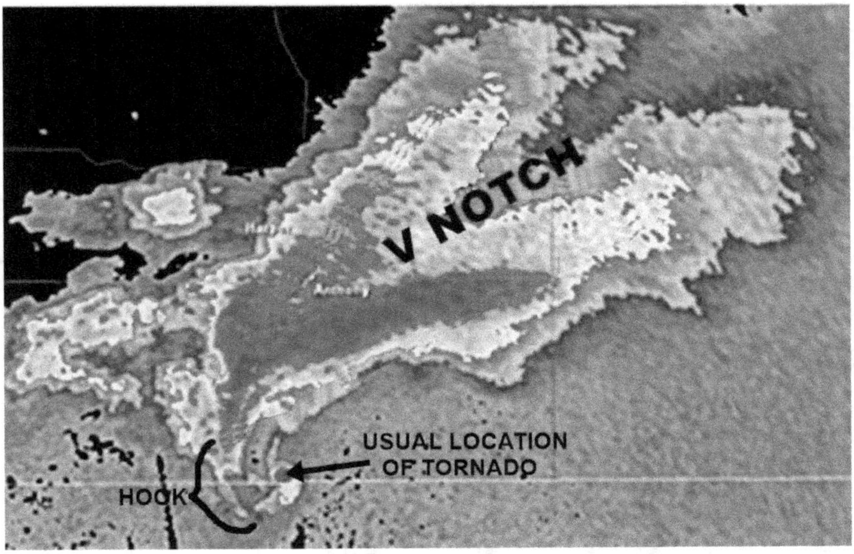

Figure 7-2 The v-notch for the Oklahoma supercell April 14, 2012. (NOAA Photograph)

supercells can be explained by the thunderstorm blocking the upper air flow, especially if it has a double vortex structure. The double vortex, as illustrated in Figure 2-5, consists of the mesocyclone that rotates cyclonically located beside a similar size anticyclonic rotation within the rain area. This is a very stable airflow pattern when the centers of the two vortices are located perpendicular to the jetstream flow. The expected location of a double vortex inside the thunderstorm that reaches to the jetstream would include the hook part of the echo representing the cyclonic rotation and an anticyclonic rotation to the right of the cyclonic rotation when viewed from the front as was shown in Figure 2-5. Without a double vortex to block the environmental air stream it is not likely that the blockage would be sufficient for a notch to be developed. Some of the rotation within both the cyclonic and anticyclonic vortices would be tilted into the horizontal by the force of the jetstream. This would cause a separation downstream creating a notch.

Doppler radar is also a useful tool for detecting a tornado. It measures the wind toward and away from the radar. When these two are together in a very small area they may indicate that a tornado is present. Some television and weather stations have these instruments.

Figure 7-3. Tornado siren.

Many communities have sirens located throughout that are activated during a tornado warning (Figure 7-3). The tornado sirens that are operated as a warning device may not cover the entire community very effectively. The sound from sirens is not likely to penetrate to people inside houses during severe weather with lightning, thunder, and high winds.

Thus, specially designed radios are available that can provide individuals with tornado warnings. The National Weather Service broadcasts weather information continuously at frequencies of 162.4 to 162.55 MHz from a large number of stations over the United States. Radios are available that receives these frequencies and can provide warnings.

Cell phones are also very useful in keeping a close watch on the weather. Apps are available from The

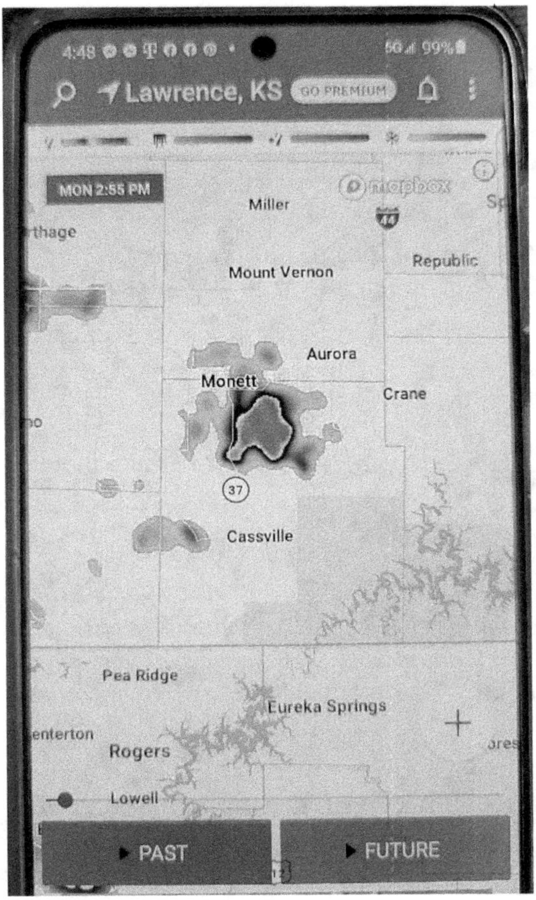

Figure 7-4. Radar is available in almost real time so that individual thunderstorms can be tracked and future locations projected.

Weather Channel and others that provide not only the forecast but radar close to real time (Figure 7-4). Once a severe thunderstorm has developed its movement can be tracked on the cell phone and its path toward or away from your location can be monitored. In addition, the hook echo may be seen if it is present.

Individual warnings may be provided by the noise of a tornado. The noise has been variously described as ten million bees, many freight trains or jet planes. The noise is so loud that it may provide a warning in itself but response time would be very short. The source of the sound is uncertain, but it could be the result of supersonic winds within the thunderstorm.

The appearance of the sky may indicate tornado formation. A "tornado sky" is shown in Figure 7-5. Such clouds form when the atmosphere is very unstable and are usually present underneath the anvil of a supercell that

Figure 7-5. A tornado sky (mammatus clouds) such as shown here indicates severe turbulence in thunderstorms and usually is present when tornadoes occur. Some thunderstorms, however, have this mammatus formation and never form a tornado.

has a tornado but their formation does not always mean that a tornado has formed.

Animals may also provide tornado warnings to people. Dogs, for example, hear sounds at ranges beyond human capabilities (the principle behind silent dog whistles). There are indications that animals can also hear tornadoes at greater distances than people. Dogs may drastically alter their behavior when a tornado is approaching. In one account, a dog that never came inside the house scratched on the door and come in when the door was opened and crawled under a bed as a tornado approached.

The chances for surviving a major tornado can be improved greatly by advance preparation and proper response during a tornado warning. Advance preparation will be considered in a later chapter dealing with building design and construction techniques. A person's immediate response to a tornado threat should be governed by a good understanding of tornado safety factors.

Tornado safety will also be covered in more detail in the next chapter but is briefly considered here. The safest location in a tornado-stricken area is within underground concrete shelters. Rooms or storm shelters constructed of at least 8 in. of reinforced concrete for the walls and ceiling are ideal shelters. Basements offer considerable protection from tornadoes if they are constructed of reinforced concrete. Damage surveys have shown that safest locations are opposite the approach direction of the tornado; thus, the northeast rooms are safer.

Safest locations in houses without basements are in the northeast rooms (rooms opposite the approaching tornado) with smaller rooms and closets safer than large rooms. Boards and debris frequently penetrate the south and west walls and these walls are also more likely to fall inward than others.

Automobiles offer little protection from a tornado if they are caught in it. Many of the deaths during the Wichita Falls tornado occurred in automobiles. The people caught in an Oklahoma tornado described in the first chapter were extremely lucky. I never heard how they got the car off the roof. A large tornado traveling at 50 mph is not as easy to maneuver around as might be expected, especially if traffic is heavy. Mobile homes are not a safe shelter during tornadoes. Mobile home parks should provide a community shelter, perhaps an underground recreational area, for use by residents.

If you are caught in open country during an approaching tornado seek a small depression such as a ditch running perpendicular to the path of the tornado. This may be helpful since winds and debris traveling at speeds of several hundred miles per hour will not make sharp bends.

We live in an age where almost everyone carries a cellphone so it is not surprising that NOAA uses a warning system that will trigger an alert on your cellphone. The Wireless Emergency Alert (WEA) procedures used by NOAA for severe thunderstorms were described in Chapter 4.

WEA alerts are also issued for tornadoes. A WEA for a tornado warning, for example, typically reads, "NWS: TORNADO WARNING in this area until 12:15PM CST. Take shelter now." You don't have to enroll or download anything to get WEAs on your cellphone, but **it does need to be enabled to receive them**.

There are a number of apps available for cellphones that are also useful for tornado alerts. **Tornado: American Red Cross** is one of the most popular tornado alert applications that may help you survive a tornado. This free application is available for iPhone and Android owners. Here you will find current information from the NOAA, as well as the ability to communicate with relatives and friends. With this application, you will hear an audible siren when NOAA issues a tornado warning. You can monitor several places that matter to you. So, it will be possible to warn distant friends and relatives too. You can track hazards on the live map. There are also step-by-step instructions (right before, during, and after) and recovery tips. Another useful feature of the app is "I'm Safe" notification sharable via social networks, text message, or email. Your relatives will know that you're safe.

Another useful and colorful application, **The Weather Channel,** as mentioned before, will help you to always be aware of the weather conditions, wherever you are. This app will always let you know the current and most accurate weather forecast. The animated main screen of the program will vary depending on the time of day, weather conditions, and location. Displaying information about temperature is possible in several ways: show a long-term forecast for 15 days or a detailed hourly one. The application also shows a lot of other weather information including radar coverage of the whole United States that can be zoomed to include details for your location. **The Weather Channel App will send you alerts when tornado warnings are issued**.

You can see a video of live coverage of a tornado warning: https://www.youtube.com/watch?v=HSFPbiyqiCk&list=RDCMUCODc8Pa-_WUMaVRp9dvmNmw&index=1

8. SEEKING SHELTER FROM A TORNADO

Tornadoes are one of the most feared natural phenomena for many people. Knowledge can replace fear but proper respect for them is justified from observations of the severe destruction that often accompanies these whirling giants. Because of the great destructive forces within a tornado and because of the difficulty in obtaining measurement in them some of their features remain a mystery. However, observations of damage after tornadoes has revealed some very consistent patterns that are repeated by tornadoes.

There are distinct categories of destruction caused by tornadoes. One type is caused by flying debris (Figure 8-1). The high-speed wind within a tornado is of sufficient strength to move small objects great distances with very high velocities, as well as large objects, even buildings, small distances. It is a matter of record that tornadoes are able to lift large objects including houses, automobiles, loaded trucks and boxcars, moving them many feet. Rock, gravel, dirt, bricks, boards, tree limbs and countless other relatively small objects are picked up by the high winds accompanying a tornado. The wind velocities in a tornado may vary from less than 100 to more than 300 mph. When a tornado is approaching, a common reaction may be to get out of smaller

Figure 8-1. Windshield penetrated by a glancing blow of a splinter of wood.

enclosures such as closets for fear of collapse. However, this could be a mistake as the flying debris can cause considerable damage and create extreme danger. Tornadoes that have been known to drive a wheat straw through a tree can certainly drive small objects through human flesh. In fact, the human skull offers little resistance to a gravel traveling at 200 mph. Even sand and soil which might seem relatively harmless can be driven with sufficient force that it can literally sandblast the skin of animals.

Observation of flying debris is helpful in understanding the winds within a tornado. In my observations of hundreds of damaged houses it has become apparent that the winds within a tornado follow a particular pattern.

For the typical tornado which may stay on the ground for fifteen miles, have a path width of three hundred yards, and travel from the southwest toward the northeast, the greatest amount of debris is flying toward the northeast. Boards, limbs, and rocks which penetrate the walls of frame houses normally penetrate south and west sides of the houses. Numerous observations have been recorded of whole houses being moved to the northeast by tornadoes coming from the southwest (Figure 8-2). This is because of two different

Figure 8-2. The southwest corner of this house fell into the basement while the northeast corner slid above ground.

factors. One of these is related to the filling of air from the back side as a tornado moves along toward the northeast. The tornado itself is composed of horizontally rotating air with an updraft core through the center of the tornado.The air in the center of the tornado has to enter at the base of the funnel and it does this from the backside as the tornado is moving along toward the northeast (Figure 8-3). Therefore, the updraft is provided by a wind from the southwest which goes into the core of the funnel. In addition, the forward speed of the tornado moving toward the northeast adds to the circular winds of the tornado so that the southeast side of a typical tornado has additional winds because of the forward speed of the tornado itself. This additional wind may be high as 60 mph.

It has also been observed from field observations of tornado damage paths as well as experiments with the laboratory tornado that a fast moving thunderstorm is likely to have a tornado with a straight damage path. The average

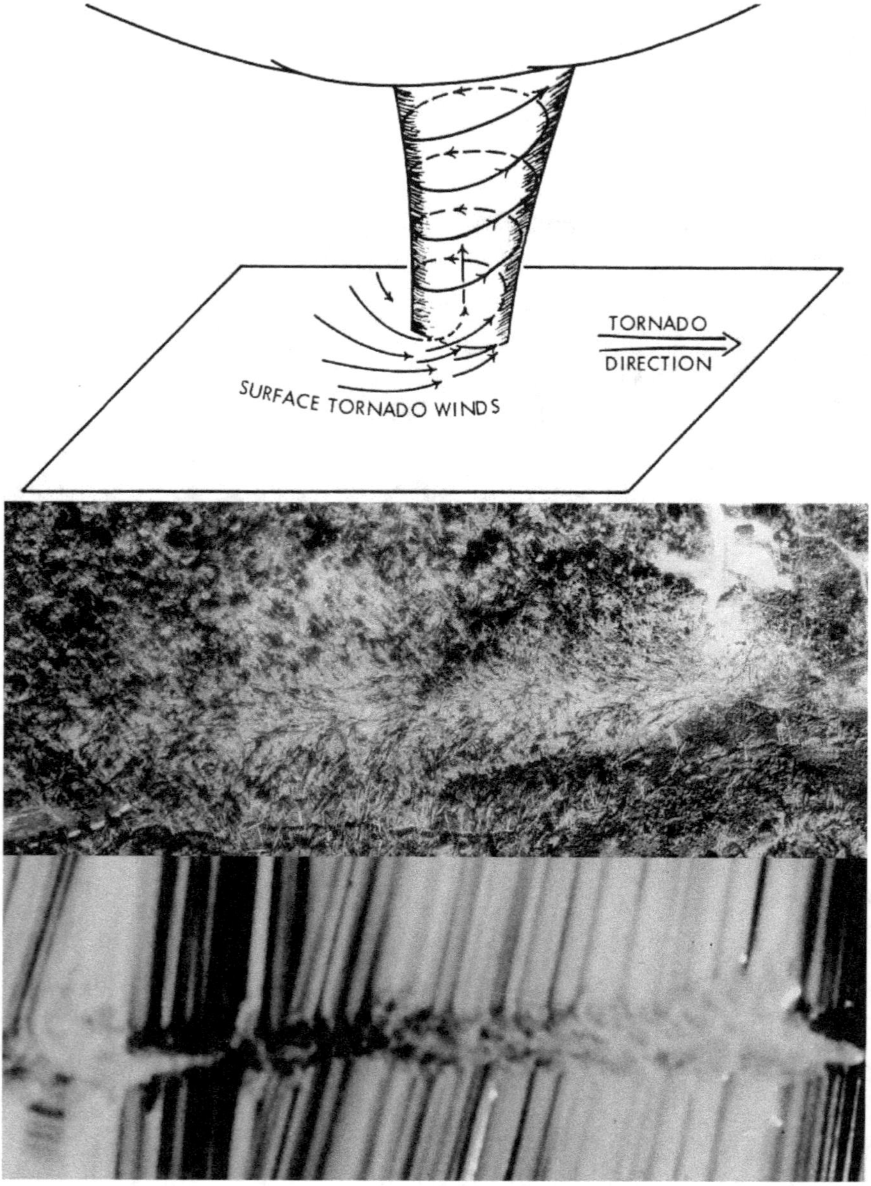

Figure 8-3. Damaging winds are concentrated in the same direction as the tornado travels because of the way the wind is pulled into the tornado. Top view is an illustration; center photograph is an aerial view of treefall directions in a forest after a tornado and the bottom photograph is from the laboratory tornado as it moved across a surface with a viscous paste put on with up-down brush strokes.

forward speed of a tornado is about 30 miles an hour. If the forward speed is considerably less than this, instead of a straight damage path, the tornado may describe a series of loops on the surface. These tornadoes have been reported by direct observation to have the appearance of an elephants trunk swaying back and forth as it approaches the observer. The path on the ground, however, would be a series of loops as it moves along.

Most tornadoes are moving with a sufficient speed so that they have a straight damage path or a slightly curved path. The typical length of a tornado damage path is fifteen miles or less, but a particular thunderstorm that has one tornado is more likely to have another tornado, since it has the right internal structure to develop them. Since about ninety percent of tornadoes come from the southwest and have a relatively straight path, the major part of the damaging winds associated with most tornadoes comes from a southwesterly direction. This is confirmed by observation of the debris carried by tornadoes which penetrates south and west walls of houses, as well as from observations of treefall directions and measurements from the laboratory tornado as shown in Figure 8-3.

The destruction of homes and buildings by tornadoes is always a great tragedy. Prior to 1980 an average of slightly more than 100 people were killed by tornadoes per year. This has dropped to an average of 70 per year since then, probably from better warnings and more appropriate responses. Deaths occur to persons being caught in the open without appropriate shelter as well as to persons seeking shelter in the wrong places within houses. For many years prior to the Topeka tornado in 1966 the southwest corner of houses was promoted by the weather service as the safest location during a tornado. However there were no damage statistics to support this location.

My research following the Topeka tornado clearly showed that the southwest part of basements and first floor of houses without basements were the most unsafe places to seek shelter from a tornado. The safest location in houses most frequently is a small room in the northeastern part of buildings. In my tornado research in many different cities I was extremely surprised at the small number of houses with basements so that many people are forced to seek shelter on the first floor. This is particularly true in the southern United States where houses are very seldom constructed with basements.

Investigations of the distributions of tornado deaths across the United States also show that there are more deaths in the southern United States than expected from the combination of number of people and number of tornadoes. A single outbreak of tornadoes on April 11, 1965 resulted in 271 deaths and injuries to more than 5000 people. This was surpassed by the tornadoes in 2011 that killed 351 people in the south.

The southwest corner of houses was heavily promoted for many decades. A survey in 1975, almost ten years after it had been debunked, conducted with hundreds of people revealed that about 70% of them would go to the southwest corner of the basement or first floor of the house for protection. The advise to go to the southwest corner still persists today but is less common.

The research that changed the tornado safety rules was my investigations after the Topeka tornado. During this tornado on June 8, 1966 about fifty people were in McVicar Chapel (Figure 8-4) on the Washburn University Campus when the tornado struck. They tried to go to the southwest corner of the chapel but in the confusion they went to the southeast corner where they escaped injury. If they had found the southwest corner they would not have been so fortunate because the large stones which composed the walls of this building caved through the wood floors and collapsed into the southwest part of this building.

Figure 8-4. McVicar Chapel after the tornado.

I talked to a lady in Lubbock, Texas after the F5 (EF5 today) tornado of May 11, 1970 who had no basement but had sought shelter in the north central part of the first floor of her house in a closet. She and her two teenage daughters felt extremely fortunate to be alive since the walls of this closet were the only parts of the house that remained after the tornado.

After the Topeka tornado, as I was looking at a large house that had been built on a concrete block foundation (Figure 8-5) the owner of the house came by and provided additional details. He said that 16 people were in the southwest corner of the basement. The whole house had been blown by the tornado about five feet toward the northeast so that boards, concrete blocks and pipes had fallen into the southwest part of this basement. The people who were in the southwest corner of this basement were covered by so much debris that they had to be dug out after the tornado but they all survived. The northeast part of the basement was

Figure 8-5 The southwest corner of this house fell downward into the basement where 16 people sought shelter.

completely free of debris since the house simply slid along the ground a few feet toward the northeast and no debris had come down in the northeast part of this basement.

These are only a few isolated cases from observations of hundreds of houses damaged by many different tornadoes, but they illustrate some typical damage patterns that it is important for people to know because they are not general public knowledge.

Investigations of hundreds of houses damaged by many other tornadoes has resulted in solid evidence that the safest place in a house during a tornado is in the part of the house opposite the approach direction of the tornado. Since most tornadoes come from the southwest, the northeast part of the house is safest most often (Figure 8-6). The second safest place is in the center of the house. The part of the house that is most damaged is usually the corner facing the approach direction of the tornado which is normally the southwest.

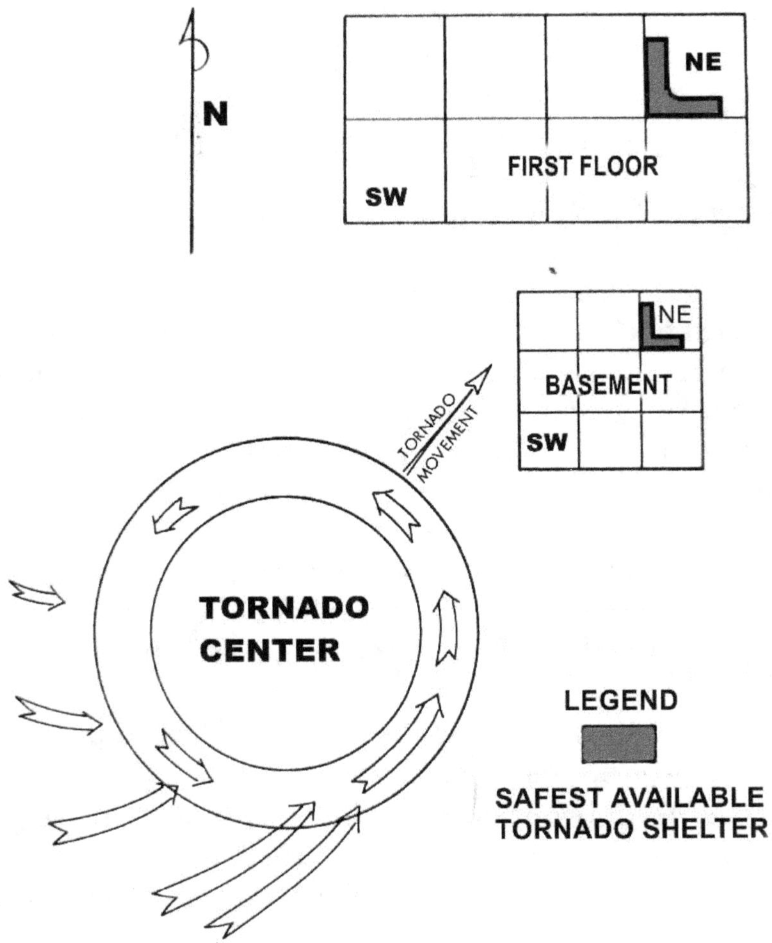

Figure 8-6. Safest locations in basements and first floor of houses.

Safest locations during a tornado listed in the order of their safety are as follows:

Indoor underground **storm shelter**. This is the safest place to be during a tornado. However, such storm shelters are rare and most people are forced to seek shelter in some other location.

Poured concrete basements with no windows. The northeast corner opposite the approach direction of the tornado is the safest location. This occurs since a part of the floor is usually left over a poured concrete foundation even with the most severe EF5 tornadoes. The part of the floor that is most likely to remain over the basement is the part downwind from the approach direction of the tornado which is normally the northeast part of the house.

Poured concrete basement with small windows. This type of basement is normally safe in the northeast part of the basement away from the windows. It is important to stay away from windows because injuries can result from flying glass and other debris coming through the windows, although it is uncommon for debris to come through north or east windows.

Half basements or walkout basements. Many modern basements are built on a slope and have only a half basement. The part of the basement that is the most submerged is the safest.

Stone, Brick or Block basements. These structures are weaker than the poured concrete basements and usually offer a higher risk of collapse. The corner opposite the approach direction of the storm is the safest because there are numerous observations of whole houses being shifted only a few feet in the direction the storm is moving allowing a lot of debris to fall into the southwest part of the basement. As a result the northeast part of houses with concrete block or stone foundations is safest most often.

The **lowest floor** of a house if no basement is available. In houses without basements the southwest part of the house is the most unsafe because of flying debris through southwest walls or walls falling inward. Do not seek shelter in this part of the house. In general small rooms such as closets or bathrooms in the northeast part of the house or the central part are the safest. Do not seek shelter against an outside wall but against inner partitions. In a two story house the lower

Figure 8-7. Automobiles are not a safe shelter in a tornado.

level offers better protection than upper levels. The best protection is provided by a small room in the northeast part of the house and the person should seek shelter against an inside wall under heavy furniture if possible.

In automobiles remember that safety is provided only if they are used to avoid the tornado (Figure 8-7). The average speed of a tornado is about 30 miles per hour. The highest speeds approach 100 miles per hour for short distances. Automobiles may offer little or no protection if they are caught in the tornado.

If a person is caught in the open country then shelter should be sought in the nearest depression by lying flat on the ground. Large trees are not safe during a thunderstorm because of the possibility of attracting lightning as well as the risk from flying limbs.

Mobile homes offer no protection during a tornado. Do not seek shelter from a tornado in a mobile home. They are particularly vulnerable to strong winds. Trailer parks should provide a community storm shelter. This could also be combined with a basement recreational facility for the trailer park residents. If no such shelter is available surrounding houses with basements would be safer or even seeking shelter underneath a bridge, in a ditch, or ravine.

In schools and office buildings the safest locations are basements or interior small rooms on the lowest floor in the northeast part of the building. Auditoriums, gymnasiums and other rooms with wide span roofs should be avoided. Stay out of hallways that run from the southwest toward the northeast and out of the southwest parts of hallways running in other directions.

Remember the hazard associated with flying debris. A gravel or straw traveling at 300 mph is not able to make sharp turns but is able to penetrate objects that it strikes. This information should be very helpful in seeking shelter from a tornado. In addition, it may be important to remember the lightning hazard and the flash flood hazard associated with severe thunderstorms.

You can see a video of a storm shelter installation at:
https://www.youtube.com/watch?v=QVrf6cSxICo

9. IMPORTANCE OF HOUSE DESIGN

Although basements are effective in providing shelter from a tornado they are available to only a small percentage of the residents in most cities. In fact, it is surprising how few basements are constructed in many localities. Basements are very uncommon in most southern cities. Even in the central United States, less than half the houses in most cities are constructed on a basement foundation.

With increasing community planning and preparation for tornado activity and with more accurate warning procedures from the National Weather Service, it is still surprising that tornadoes do not kill more people each year. One of the reasons for this may be the over-pressure inside houses as a tornado moves by, creating a tendency for the roof to go up and the walls to go outward for all except the south and west walls that are blasted by very high velocity winds coming from the southwest.

In addition to the knowledge of safest locations in houses we could do a great deal to decrease the damaging effects of tornadoes by constructing houses that are safer. The problem consists of providing adequately designed and constructed houses that resist higher wind velocities. With increasing costs of building materials many contractors and developers cut corners to retain profit margins. This frequently results in a house which is much less able to withstand the high winds associated with a severe thunderstorm or a tornado. One concept is that tornadoes are so powerful that they are going to destroy any kind of building and, therefore, we should not even be concerned with building codes or attempts to design and construct houses that can withstand high wind velocities. Tornadoes come in a wide range of intensities and houses can be designed to greatly increase their chances of surviving the winds of most tornadoes. Most tornadoes (95%) have wind speeds less than 135 mph and houses can certainly be designed to withstand these velocities.

Our research at the University of Kansas using model houses in a wind tunnel has shown houses can be designed to withstand higher velocity winds. Primary factors in designing and constructing wind resistant houses consist of paying attention to the orientation of the house, the slope of the roof, the strength of connections and reinforcement techniques.

The pressure patterns over a house subjected to high wind velocities were measured. This has shown that the orientation of the house is very important in its ability to withstand high wind velocities. For a rectangular house the short end of the house should be turned toward the west or southwest for a low roof angle and toward the south for a steeper angle. This gives greater resistance to the high wind velocities that can be expected in a tornado and would allow a house to withstand the winds of a large number of tornadoes.

The slope of the roof has been found to be very important in withstanding high wind velocities. Low angle roofs fly because of the pressure distributions exactly the same way that an airplane wing supports the weight of the aircraft; therefore, flat or low angle roofs are to be avoided, if possible. Steep angle roofs of 30 or even 45° do not suffer from this disadvantage and are able to withstand higher winds. Therefore such simple factors as the orientation of a house

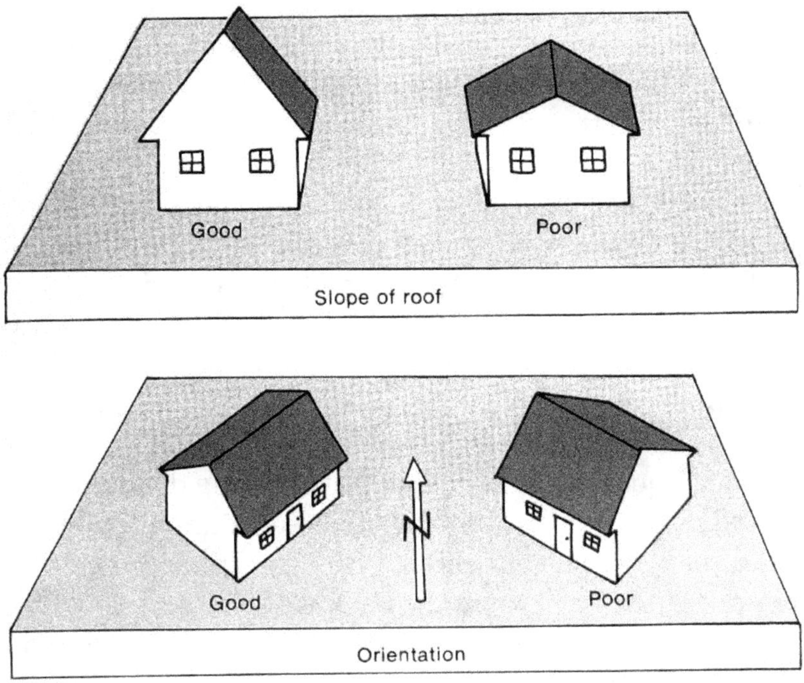

Figure 9-1. The design of a house contributes to its ability to withstand strong winds. Steeper slopes and houses with the short end oriented toward the southwest are more likely to withstand high winds.

and the selection of roof angles are important factors in designing for higher wind velocities.

Major weaknesses in frame houses may be found at the connections of roofs to walls and walls to foundations. If these connections can be made as strong as the wood members themselves, it is possible to construct houses to withstand wind velocities of at least 150 mph and this covers the majority of all tornadoes. The strength of the connections can be increased by simply using more nails, but a better approach would be to use galvanized steel clip framing anchors (hurricane clips) at every roof to wall connection and at the wall to foundation connection for those parts that are not bolted to the foundation. Additional bolts placed in the foundation so that the wood members of the frame house are bolted securely would solve the problem of weak foundation to wall connections. A three or four foot placing of bolts would be sufficient to secure the house to the foundation.

Current construction practices are much less rigid. The roof may be held to the wall by only a few nails assuming that the weight would support it. The same assumption is made for the foundation to wall connection.

Observations of tornado damaged houses that were constructed on poured reinforced concrete foundations frequently revealed that even in strong tornadoes, if the roof and walls were blown away the floors remained over the poured concrete foundation when this connection was strong. I saw some houses, however, that did not have the nuts on the bolts that were anchored in the concrete foundation so that the house was simply lifted off the poured concrete foundation and set down in the back yard.

Building materials that use mortar are usually very weak. Concrete block foundations are generally very weak but can be made quite strong by the proper reinforcement techniques. When blocks are held together only by cement between them, little resistance to a tornado is offered. When long metal rods are inserted down through the empty spaces inside the concrete blocks and secured with cement inside the blocks they are connected together. This makes them quite strong and able to withstand high velocity winds. It is not possible to reinforce stone or brick foundations enough to allow them to withstand the winds generated by a tornado.

We also investigated the effects of appendages on a house, such as porches and chimneys, on wind damage. Models were prepared and tested with results that showed such modifications were helpful in reducing damage. The airflow was modified by the obstructions in a beneficial way for all roof angles.

Another test involved differences between one- and two-story houses. This showed only minor differences between the pressure over the roof for each of the two designs if they had the same roof angles. But for each design the upper floor of the two-story house was damaged sooner than the first floor of the one-story house.

The patterns of failure were also of interest. The first parts of a house to be damaged as the winds increased were the eaves of the roof on the upwind side of the house and the downwind side of the peak of the roof. Major structural damages had occurred by the time the winds reached 150 mph. As part of the roof blew away, the potential for additional structural damage increased. A model building destruction sequence is shown in Figure 9-2.

In summary, wind damage is most likely to occur differentially according to roof angle and orientation. Steep roofs are safer in strong winds than low roof angles. Wind

Figure 9-2. This series of photographs shows progressive damage during a destruction model test. Damage occurred first at the peak of the roof, then to the roof itself. Wind is blowing from left to right.

damage is most likely at the upwind edges and along the peaks, at the roof-wall joint or the wall-foundation joint. The downwind part of a house is likely to receive less damage than the upwind section.

The destruction model testing also showed that the southwest part of the first floor was unsafe because of the greater chance of structural damage in this part of the house, in addition to the greater chance for penetration of debris into this area. For houses with weak foundations, the southwest part is unsafe because of the tendency of tornadic winds to move the house off its foundation with displacement toward the northeast. Houses with good foundations (poured reinforced concrete) are also the most unsafe in the southwest part because of the increased risk of the house and ceiling over the basement being removed from over the southwest room before those in other locations.

Houses with vents were tested to determine the effect of ventilation on damage. In general, vents in locations comparable to doors and windows on a single story house caused it to receive greater damage if venting was provided on either the upwind or downwind side. However, vents on the downwind side of the roof near the peak allowed the house to withstand stronger winds. In terms of designing houses that would correspond to placing vents of the north and east sides of the roof.

A number of applications can be made that will allow houses to withstand stronger winds. In addition to the importance of slope and orientation of the house, the proper strengthening of joints will allow a house to survive much stronger winds. Strengthening of the roof to walls by hurricane clips has already been mentioned. The walls should be securely anchored to the concrete foundation with bolts embedded in the concrete. Venting of the roof should be provided near the peak on the north or east side.

This can serve the added purpose of releasing hot air from the attic during the summer.

While houses with poured reinforced concrete foundations with good wall to foundation connections are generally safe shelters from a tornado, additional security is provided by constructing a storm shelter as a part of the basement. Such an indoor shelter is preferable to an outdoor underground shelter because of the danger in getting to it during the short warning time for most tornadoes. The shelter should consist of concrete walls at least eight inches thick with a top slab of eight inches of concrete reinforced with metal bars connecting the walls and ceiling at nine inch intervals. A concrete hallway with a 90° turn or strong door should form the entrance. This should prevent flying debris from entering the shelter. If such a shelter is constructed when the house is built it represents a very small additional investment.

You can see a video of building tornado proof houses at: https://www.youtube.com/watch?v=ga7N-bS4nLU

10. CONCLUSIONS

Tornadoes form from thunderstorms that have the right internal structure. Such thunderstorms are developed under specific atmospheric conditions. They usually develop in the warm air sector of a frontal cyclone. Their usual direction of travel is from the southwest. Other damaging features of a severe thunderstorm may include the presence of lightning, hail and microbursts.

Misconceptions can potentially be very dangerous, such as the notion that the southwest part of a house is the safest location. Investigations show that this is the worst place for seeking protection during a tornado, and yet it is the location where some people still seek shelter during a tornado.

Another misconception is that we can't design buildings to withstand tornadic wind velocities. Evidence indicates that an ordinary frame house can be designed to withstand winds of 135 mph or more, that would allow it to withstand all of the most frequent tornadoes. Other considerations like building a house with a steeper roof, such as a 45° roof, allows it to withstand higher wind velocities than a low angle roof. A house with the proper orientation also has a distinct advantage in high winds. For low-angle roofs the proper orientation is with the short end of a rectangular house turned toward the west or southwest. For a house with a steep roof angle the orientation is not as critical, but has some advantage in high winds if the short end is turned toward the south or southwest.

The best protection from a tornado is provided by indoor underground shelters constructed of reinforced concrete. If these are not available basements constructed with reinforced concrete provide very effective shelter from tornadoes in areas away from windows that may be located in the above ground portions of the basements. The northeast part of reinforced concrete basements should be chosen for protection, since the floor frequently remains above this part of the basement and is frequently removed from over the southwest part of basements.

For basements constructed of stone or concrete block the northeast part of the basement offers the best protection because of the tendency of the blocks or stones to crumble allowing debris to fall into the southwest part of the basement. The whole house may also be shifted some distance toward the northeast resulting in greater protection in the northeast part of the basement than in the southwest. In two-story houses without basements the lower floor is obviously the better choice for seeking protection from a tornado.

Proper appreciation of the lightning hazard is also important. A general rule to remember is that lightning is most likely to strike the tallest objects during a thunderstorm. There may even be a warning before lightning strikes so this may be important to recognize. If you feel an electrical charge building up that causes your hair to stand on end or your skin to tingle, lightning may be about to strike you. If you drop to the ground immediately the lightning may strike another location. It is also important to know that a person apparently killed by lightning can often be revived by prompt cardiac message, mouth to mouth resuscitation, and prolonged artificial respiration. Sharp blows by the open hand should be administrated to the chest of the victim in an attempt to start the heart to functioning again. This should be followed by mouth to mouth respiration. While it is important for this type of information to be readily available it is more important for emphasis to be placed on prevention of lightning and tornado victims. This can be accomplished only by the use of appropriate judgment during severe weather in addition to appropriate prior planning through the design and construction of your home.

I hope the information you have gained from this book proves useful to you and improves your ability to deal with severe weather. If you are interested in further information including more details on tornadoes and severe weather you are referred to the expanded coverage in my related book published in 2021: ***Understanding Severe and Unusual Weather*** along with it's workbook.

Further Reading Choices

ABOUT THE AUTHOR

Joe R. Eagleman (1936-) was born on a farm near West Plains Missouri. He received the PhD from the University of Missouri in 1963 and served as professor at the University of Kansas for 39 years and continues as Professor Emeritus. He taught thousands of students about Atmospheric Science through his courses there and many thousands more through four different textbooks used by over a hundred universities over a span of several decades.

He directed a successful experiment on Skylab, funded by NASA, and invented a tornado in his laboratory that was used by Universal Studios for a 50 ft. tornado attraction in the Twister Building in Orlando Florida for several decades and can still be seen at the Exploratium. He is also the author of a technical book on severe thunderstorms that includes his tornado safety research which resulted in changes that were adopted nationally.

His autobiography, *Name Your Price*, tells of his early life on a farm where he was the 11th of 12 children. It includes his work as a professor and scientist as well as a number of unusual hobbies including those as an artist, musician, luthier, marksman, taxidermist and world traveler.

Since his retirement he has recorded five albums of original music. His last two albums were released in 2019 and contain 24 of his original songs. He has also written several books since his retirement.

For more information see http://www.JoeEagleman.com.

www.ingramcontent.com/pod-product-compliance
Lightning Source LLC
LaVergne TN
LVHW011816200926
848484LV00005B/1217
* 9 7 9 8 2 1 8 7 0 8 3 4 4 *